HORMONES OF INSECTS

RECENT ADVANCES IN ENTOMOLOGY SERIES—5

HORMONES OF INSECTS

Edited by

A. B. Saxena

ANMOL PUBLICATIONS PVT LTD

NEW DELHI - 110002 (INDIA)

ANMOL PUBLICATIONS PVT LTD
4374/4B, Ansari Road, Daryaganj,
New Delhi - 110002

Harmones of Insects

First Edition, 1996
Reprinted, 2000
ISBN 81-7488-350-9

PRINTED IN INDIA

Published by J. L. Kumar for Anmol Publications Pvt Ltd, New Delhi, and Printed at Mehra Offset Press, Delhi.

Preface

This fascinating title has been carefully organised to meet the long-felt needs of increasingly large number of those who deals with different aspects of Entomology. It provides a balanced and integrated treatment of the entire field of entomology. The title is intelligible to the educated laymen but it deals with some complex ideas. It is an adequate text for all requirements in this area for most university students. This ground work with its many new and unfamiliar names is, at any time, dry, uninteresting and difficult, especially as the beginner does not at first see the necessity for it. We have endeavoured to make it interesting by introducing as much comparative morphology and phylogeny, as possible in the limited space at our disposal.

The primary aim throughout has been clarity so as to make the text selective and meaningful. We hope also that general Entomologists and students of Agricultural Entomology will find in this book valuable and recent informations about the subject and also allied branch of the subject. The book provides a nucleus around which the teacher can plan a successful course without much bearing on complementary and supplementary readings. The theme of the book is novel and for the first time the various chapters in different volumes have been organised reviewing the available literature. All the topics have been illustrated with line diagrams, graphs and tables wherever necessary.

We hope that the book will definitely prove to be a boon to teachers, students and research scholars.

Author

Contents

1

The Protohormones

One of the most characteristic features of hormones is that they produce their effects in parts of the body other than those from which they are secreted, and reach their site of action by means of the body fluid. This should, however, be regarded as phylogenetically secondary and the result of a long and gradual process of evolution. A part from the various true insect hormones so far considered, there have been found in arthropods a series of physiologically active substances which have their site of action at the place where they are produced and which are only occasionally carried in the body fluid when they may exert their effects elsewhere.

These substances no doubt represent a stage through which all the present hormones in every animal group had to pass in their phylogeny. It therefore seems useful to classify them as a special group of physiologically active substances, giving them the term protohormones or prehormones. They may be regarded. as hormones in *statu nascendi*. Most of the present protohormones are substances whose evolution has stopped at this level, and which carry out their particular functions in the body of present-day animals, just as, for example, rudimentary light- sensitive spots instead of complex eyes are found contemaporaneously with the very complicated compound eyes of arthropods, or as the primitive.protozoans appear alongide the most complicated and phylogenetically advanced vertebrates.

The known protohormones may be divided into two groups: *the neurohumoral factors* which have been characterized together with the neurohormones in the previous chapter, and *the gene hormones*, which may be defined as substances which depend on particular genes and which condition particular genetic features, but can produce their effect by means of the circulation of the body fluid, even when introduced into animals which lack the appropriate genes.

NEUROHUMORAL FACTORS

The most widely occurring and the earliest known type of secretory products from nerve cells are the so-called neurohumours, i. e., physiologically active substances such as acetylcholine and noradrenaline in the nerve endings. The definition of hormones applies to these substances in that they are produced by specific cells, they are physiologically active and they produce their effect inside the organism. The definition does not apply, however, in that the substances work at their site of origin (in the nerve endings and synapses) without entering the body fluid under natural conditions. They thus correspond to the concept of protohormones as used in this book. Some of them are closely related, if to identical, with some of the neurohormones in their chemical composition. Also most, if not all, of the neurohormones will have probably passed through such a stage in the course of their phylogenetic development. In some instances it is difficult to decide whether a particular active substances isolated from nervous tissue (e. g. acetylcholine) has the character of a neurohumoral factor or perhaps that of a neurohormone produced and discharged in the form of neurosecretion.

The function of most of the known neurohumoral factors is their share in the transmission of nervous excitation in the synapses. This seems to apply to acetylcholine, which is known from the nerve endings (synapses) of the parasympathetic nerves and the preganglionic parts of the sympathetic nerves of vertebrates. Bykown and Kibjakov have demonstrated that acetylcholine is also present in nerve endings in the central nervous system. In the postganglionic parts of the sympathetic nerves, the function of acetylcholine is replaced by that of noradrenaline (sympathin). The reactions in the synapse can be regarded as consisting of the following stages; the electric current reaching the synapses from nerve processes starts specific chemical reactions. These result in acetylcholine being liberated, which produces a similar change in the electric potential in the nerve fibre (axon or dendrite) of the next nerve cell which, in turn, is transmitted to the other pole of this nerve cell in the form of an electric current.

Similar, but more complicated, changes take place through the action of other neurohumours at the neuromuscular junctions of other nerve endings, resulting in stimulation of the muscular or glandular tissue concerned or in the induction of excitement (centripetal electric current) in sensory endings. The following neurohumoral factors have been found in insects.

Acetylcholine

This has occasionally been found in very high concentrations in insects, e. g. in similar weights of tissue, up to fifteen times more acetylcholine was found in the nervous system of the cockroach *periplaneta americana* than in the central system of vertebrates (Mikalonis, 1941; Tobias, 1948). The question remains whether acetylcholine occurs not only in the nerve endings but also as a true neurohormone inside the neurosecretory cells. Besides acetylcholine, acetycholinesterase has also been found in large quantities within insects, again in *Periplaneta americana* (Richards and Cutkomp, 1945) and also in *Pis mellifica* and *Melanoplus differentialis* (Means, 1942).

Acetylocholinesterase

This is most abundant in the neuropile, where most synapses occur. It is also present in the neural lamella of each ganglion and in the glial membranes, where its function seems to be to protect the central nervous system from the entry of acetylcholine from outside (Treherene and Smith, 1965). Its first appearance in insect embryos was found to concur with the appearance of neuroblasts (Treherene, 1966). On the other hand, as distinct from the condition in vertebrates it is absent from neuromuscular junctions, suggesting that, in insects, it is not the chemical transmitter in this type of nerve ending.

Serotonin (5-hydroxytryptamine)

This has also been found in considerable amounts in insect ganglia and in corpora cardiaca (Treherene *et al.*, 1967). It seems to be produced by specific neurosecretory cells (Hinks, 1967) and to have various excitatory and inhibitory effects in different tissues and species, e. g., stimulation of the heart rate and gut peristalsis in *Periplaneta*, stimulation or inhibition of neuromuscular transmission and enhancement of night inhibition of neuromuscular transmission and enhancement of night flight in Lepidopytera, stimulation of secretory activity in isolated salivary glands of *Calliphora*, stimulation of the action of the diuretic hormone in *Rhodnius*, etc. (cf. Wigglesworth, 1970).

Hirpi and Salanki-Rosza (1973) studied the presence of 5-hydrotryptamine (5HT) and catecholamines in the central nervous system and heart of *Locusta migratoria* with an Aminco Bowman spectrophotoflurimeter, according to their excitation and emission spectra. 5HT and dopamine were found in considerable amounts in both the gangiia and the heart.

Noradrenaline was present in the cerebral ganglion only, and in only about 10 per cent of the dopamine concentration. Adrenaline was found neither in the ganglia nor in the heart. The three positive substances are regarded as transmitter or modulator substances in the insect CNS and 5HT and dopamine in the heart also.

Catecholamines

Catecholamines have been found in insects in relatively large amounts. Dopamine (3-hydroxytyramine) seems to be the most abundant, while adrenaline and noradrenaline occur in smaller amounts (Treherene, 1966). Among their various important effect, adrenaline, noradrenaline and dopamine were shown to stimulate the activity of the central nervous system, followed by blocking of the giant fibre synapses, in cockroaches (Hodgson and Wright, 1963). They produce pharmacological effects on the heart rate, gut peristalsis, the Malpighian tubules and the oviducts (Wigglesworth, 1970). An adrenaline-like transmitter affects the luminous organ in the fire-fly *Photuris*, where the application of reserpine (known to counteract adreergic transmitters in vertebrate nerve endings) abolishes the light produced by nerve stimulation. Catecholamines can be demonstrated histologically by the characteristic yellow-green fluroescence in lyophilized ganglia, specific areas in the cerebral ganglion of *Periplaneta* (Frontali and Norberg, 1966) and the ventral nerve cord of *Gryllus* and other insects (Cech and Knoz, 1971).

GABA (Y-aminobutyric acid)

This is another neurohumoral factor common to insects. It has been shown to inhibit impulse conduction in caterpillars of

Dendrolimus pini. It is formed by decarboxylation of glutamic acid. The corresponding decarboxylase is also of common occurrence in the insect nervous system (Treherene, 1966). L- glutamate was shown to induce depolarization in the synapses in this way (Usherwood, 1968).

Certain other pharmacologically active substances

Such substances have been found in insect tissues, but their chemical structure and possible function in the insect organism have not yet been determined. A substance which is produced in the opaque male accessory glands of male insects, and stimulates the nerves controlling peristalsis in female reproductive tracts, merits special attention. It seems to be a pharmacologically active amine which stimulates the heart rate, similar to the one produced by the pericardial cells in *Periplaneta* through the action of a peptide neurohormone from the corpora cardiaca (Davey, 1964; Wigglesworth, 1970)

GENE HORMONES

Gene hormones have so far been demonstrated only in insects, although they can be presumed to exist in other groups of animals also (at least in other arthropods). They are clearly of a protohormone character. The messenger RNA responsible for induction of the enzyme synthesis mentioned below can probably be regarded as being the actual gene hormone. Thus the term 'gene hormone' is well justified. This also follows from the existence of morphogenetic gene hormones as mentioned further below. At present, the following types of gene hormones are known.

Gene hormones and eye-colour of *Ephestia kuhniella*

Perhaps the first gene hormone discovered was the one which conditions the development of the black colour in the compound eye of *Ephestia kuhniella* (Caspari, 1933; Plagge, 1936a, b. A red-eyed mutation has been isolated in cultures of this moth, whereas the eyes of the wild type are black. Genetical analysis has shown that red eye-colour is recessive to black. The wild form AA is further characterized by dark pigment spots on the testes and black pigmentation of the larval epidermis and of the ganglia of the nervous system. These features are all missing in the aa form.

It has been shown that the colouration peculiar to the wild form AA may equally develop in the red-eyed mutant if testes from normal males are implanted during a particular sensitive period before pupation. Transplantation of the ovaries, the brain and even of individual eggs from the AA form, produce the same effect as transplanting the testes. This effect is not species specific; implantation of ovaries from the moth *Acidalia virgulata* and transfusion of haemolymph from pupae of *Galleria mellonella*, and even from the flies *Drosophila melangaster* and *Calliphora erythrocephala* have the same effect (cf. Plagge, 1936a, b; Ephrussi, 1942a, b). It is interesting to note that the effect is not directly proportaional to the amount of gene A. It has been shown that the tissues of the hybrid race Aa have the same effect as those of the AA form. From this it is concluded that the effect is not due to the gene itself but to a substance produced in the tissue when the gene is present. It has further been shown by Plagge that the presence of AA tissue for 24 hours is sufficient to produce the effect whether the implant is then removed or not. A possible explanation is that the pigment may catalyse its own further synthesis once it has developed in the new medium (the aa host).

Gene hormones in the eye-colour of *Drosphila melanogaster*

Even more detailed investigations have been carried out on the development of eye-pigmentation in *Drosphila*, where a similar proto-hormone effect has been demonstrated. Genetical experiments with *Drosophila* have produced numeorus mutations differing in slight degrees of tone in their eye- colouration, from the dark red of the wild type to pure white. Some of these colourations are effected autonomously by genes, while in others a hormonal effect very similar to that described for *Ephestia kuhniella* has been found. The difference between these two groups of colourations is evident in gynandromorph individuals of *Drosphila* from the fact that, whereas in the former group each eye can have a different colour tone, in the latter group, if one eye is of the wild type (w), the other necessarily shows the same colouration. Of the many mutations described the following have been shown to be hormonally controlled: vermilion (v^+), cinnabar (cn^+), cardinal (cd^+) and scarlet (st^+). The substances controlling these colour tones are again present in various parts of the body such as the Malpighian tubules and the fat body. By transplanting these tissues the particular eye-colour may be induced in the mutants lacking the relevant gene. Similarly, if an imaginal eye disc with a particular gene missing is transplanted into a larva which has that particular gene, the eye which develops from the disc shares the colour tone of the host. It has been shown that these gene hormones produce their effects even when they are administered in food, though of course only if accepted by the larvae during the short period of about 70 to 65 hours before pupation (Beadle and Law, 1938).

The chemistry of the gene hormones

Much attention has been paid to the chemical composition of the gene hormones which affect eye-colouration. The careful investigations of Butenandt, Weidel and Becker (1944) and Tatum and Beadle (1938) have shown that the gene hormone A of *Ephestia kuhniella* and the gene hormone V^+ of *Drosophila* are identical. Moreover, their effect can be repeated with 1-Kynurenine, which has some physical and chemical features in common with the gene hormones. In contrast, d- Kynurenine and kynurenic acid have no effect on eye-colouration.

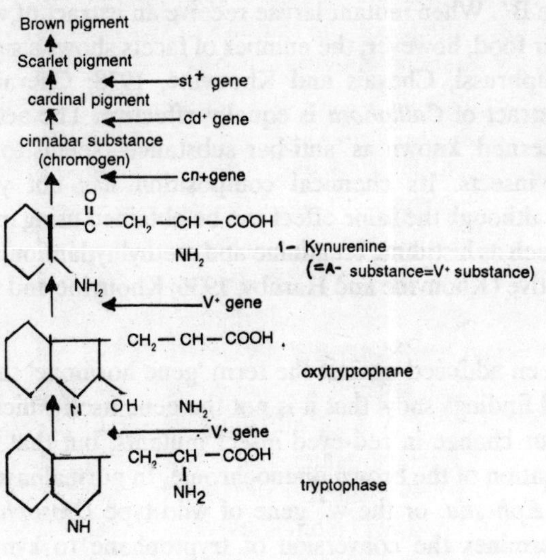

L-kynurenine is a substance formed by the enzymatic oxidation of the amino acid tryptophane. The gene hormones concerned have been shown to work as oxidation enzymes in such a way as to continue the reaction until the particular colour substance has been formed. Table 1 shows how, according to Beadle (cf. Schjarrer, B., 1942), the various eye-pigments develop.

The brown eye-pigment of *Drosophila* is a compound of the ommochrome group (Becker, 1942). The cinnabar pigment (cn^+) may be regarded as the chromogen for the black eye-pigment (Kikkawa, 1941). Quantitative investigations on the formation of eye- pigment in *Ephestia* have shown that the amount of pigment is directly proportional to the amount of the injected pigment.

Other gene hormones

Apart from the gene hormones affecting the pigmentation of the eyes and various other parts of the body, a humorally transmissible factor which conditions the number of facets of the eye has been shown in *Drosophila*. In *Drosophila melangaster* a mutation known as 'bar' occurs, which differs from the wild type in that its compound eyes have 50 to 60 facets instead of the normal 600 to 650 facets. It is said to lack the gene B^+. When mutant larvae receive an extract of wild type larvae in their food, however, the number of facets shows a significant increase (Euphrussi, Chevais and Khouvine, 1938; Chevais, 1943, 1944). An extract of *Calliphora* is equally effective. The active substances concerned, known as 'anti-bar substance', seems to be very common in insects. Its chemical composition has not yet been determined, although the same effect can be obtained using imidazole derivatives such as histidine, creatinine and methylhydantion, which is especially active (Khouvine and Harnby, 1936; Khouvine and Chevais, 1938).

It has been adduced against the term 'gene hormone' that some experimental findings show that it is not the gene itself which causes the eye-colour change in red-eyed insect mutants, but that is is the inhibits formation of the brown ommochrome. In normal insects (the AA gene in *Ephestia*, or the w^+ gene of wild type *Drosophila*), the enzyme determines the conversion of tryptophane to kynurenine, which is then transformed to hydroxykynurenine and further to the brown ommochrome (cf. Wigglesworth, 1970). The enzyme is evidently a product of protein synthesis determined by the relevant DNA locus, so that the term 'gene-dependent' hormone is still justified. The substance produced in transplantation experiments is undoubtedly the corresponding mRNA. Findings of morphogenetic protohormones of this type testify in favour of this conclusion. As mentioned by Wigglesworth (see above), the findings on insect gene hormones were one of the first and clearest examples in support of the 'one gene'–one enzyme' hypothesis generally accepted in modern molecular biology.

2

The Exohormones

The exhormones can be viewed as hormones for they are composed by the organism and are physiologically active at extremely small concentrations. Nevertheless, they do not work alone consequently (if at all), in the individual which produces them, but inplace affect other individuals in the same colony of social insects. Some authors have called them 'socio-hormones'. The term 'ectohormones' has also been applied, but this is improper because not all exohormones occur on the body surface of the individuals producing them and they are rarely created or influence there produced there or act there.

The title suggested by Karlson and Lüscher (1959a, b) is 'pheromones'* for this kind of active substance, the term to include substances 'secreted by an individual and received by a second individual of the same species in which they release a specific reaction, for example, a definite behaviour or developmental process. The principal of minute amounts being effective holds. They perform as messengers between chemicals.

Besides the previously defined exohormones the authors also include in the term pheromones (later changed to pheromones by Pain, 1961) sexual attractants and other olfactorily active substances which reach their site of action through the air, such as marking scents and warning substances. The other orally active substances are induced by the authors such as the alleged determining substances in royal jelly.

In a different research work on the same class of substances Karlson and Butenandt (1959) suggest the 'possible future necessity'

*From the Greek 'pherein' = to carry, and 'hormon' = to stimulate, to excite.

of distinguishing between substances which act sensorily, for which the term telomones is suggested, but the ones which act biochemically, the term pheromones is suitable.

We cannot encompass both the sensorily and biochemically active substances under the same title, within which one may distinghish the biochemically active substances working inside the body (hormones) from those working outside the body (exohormones). Firstly, a basic phylogenetic difference exists between the two groups. Those substances which act sensorily, indirectly and by means of the nervous system have secondarily acquired a physiological action on a particular function (e.g. copulatory movements) after that function had been fully developed phylogenetically. By contrast those substances which act biochemically and directly have an activity which is phylogenetically primary. From this a more complicated function (e.g. inhibition of ovary development by the queen inhibiting substance of the honey bee, or inhibition of morphogenesis in permanent nymphs by inhibitory substances of the reproductives in *Kalotermes*) has developed. This clear demarcation between the exohormones and substances acting sensorily exists even though both types of substance seem to work together in some cases as in the control of honeybee workers by the queen. The distinction furthermore holds even though there may exist a group of substances which control neurosecretion and which seem to combine both an indirect (sensory) and a direct (biochemical) effect but they are however of a clear-cut sensory character.

The additional indications show that there is a close relationship between exohormones and hormones in the restricted sense is provided by the extremely interesting effects that exohormones of the honey bee have on the development of the ovaries in crustaceans and particular neurohormones are resemble their physiological effects.

The basic connection of exohormones with hormones is even explicit if we view the social insect colony as the beginnings of an organism of higher order. Therefore we can consider exohormones as substances circulating and producing their effects inside the organism of the higher ranking, i. e. the colony.

Therefore, it is desirable to limit the title pheromones to those substances which act via the nervous system (the sense organs) and to reserve the term exohormones for substances with a direct biochemical activity being transmitted within the colonies of social insects. The previous editions of this book have used this term with similar implications.

Currently, Lüscher (1972) has put forward that the metamorphosis hormones, MH and JH, together with JHa, may also act as exohormones (pheromones in his terminology, but with a different meaning from the one given to this term in the present book). His conviction is based upon the discovery that the hormones can be passed from individual to individual by trophallaxis, by faeces and by vapours. Confirmation of this is provided by the experiments of Springhetti (1969, 1970), which proves that reproductives, with their excess of JH, stimulated the production of soldiers and inhibited the production of further reproductives – can be assigned to (and was actually demonstrated experimentally as being due to) JH action.

The active substances which are contained in royal jelly can be regarded as a special group of 'milk' (or nourishing) substances which are neither pheromones or exohormones as they work mainly nutritionally. In this connection the term 'milk' is used in its widest

TABLE 1

Physiologically active substances

	enzymes	digestive respiratory etc., etc.
directly (biochemically) acting substances	hormones	hormones s. str. protohormones exohormones
	vitamins	vitamins s. str. determining substances (?) milk substances
Indirectly (sensorily) acting substances (= pheromones)	neurosecretority active scents(?)	inhibitory odours phase determining odours(?) stimulating odours(?) behavior influencing odours
	telomones	sex attractants making scents territorial demarcation scents warning scents

physiological sense to include the pharyngeal products of bees, i.e., the royal jelly and worker food as well as the milk of mammals. The classifications of the physiologically active substances is depicted in Table 1.

However only hormones and exohormones are discussed in this book. But few of the pheromones of bees and termites have also been included, as they are produced together with the exohormones and till now they are not distinctly individualised.

THE EXOHORMONES OF TERMITES

Polymorphism

Notwithstanding the low phylogenetic level of the termites, the highest level of caste polymorphism and division of labour is existent in this group. The dissimilarity of types shows a considerable advance over the social Hymenoptera with the occurrence of termites of all the castes. Not only reproductives, but also workers and soldiers of various kinds are produced in both sexes. All this is undoubtedly associated with a much higher phylogenetic age of the polymorphism here than in the phylogenetically younger Hymenoptera.

Normally a colony of fermite has one pair of reproductives, the king and queen, which were initially winged and only lose their wings during the foundation of a new colony after the mating flight. The balance of workers and soldiers in a normal well-developed colony of a particular termite species somewhat constant. In many phylogenetically primitive termite species, however, there are no true adult workers, and their position and function is occupied by the so-called pseudergates. These are larvae whose development has been arrested at a particular morphogenetic stage prior to metamorphosis. Their prothoracic glands remain intact and active, however, so that they continue moulting at regular intervals, but without any growth of differentiation. These moults have been called stationary moults. The pseudergates are therefore permanent larvae in the full sense of the word. When the colony is deprived of the reproductive pair the development of the pseudergates may carry on to moult and consequently form supplementary reproductives.

The larvae whic are hatched in *Kalotermes* and other phylogenetically primitive termite species are indentical, without any morphological or potential caste differences. Until the end of the IIIrd larval instar each nymph in the colony may eventually become a reproductive, a soldier or a worker (pseudergate). During the IVth instar some of the larvae develop0 to become soldier and later which after next moult becomes a soldier which becomes a soldier after the next moult. Following the fifth instar speciman may become either a pseudergate with stationary moults or it may moult to a supplementary reproductive or to a pre-soldier. Otherwise, it may carry on to assume the adult form, multing to the Ist and IInd nymphal instars

TABLE 3

Polymorphism in Kalotermes flavicollis. Each arrow represents a moult, the broken line indicates the possibility of several moults.

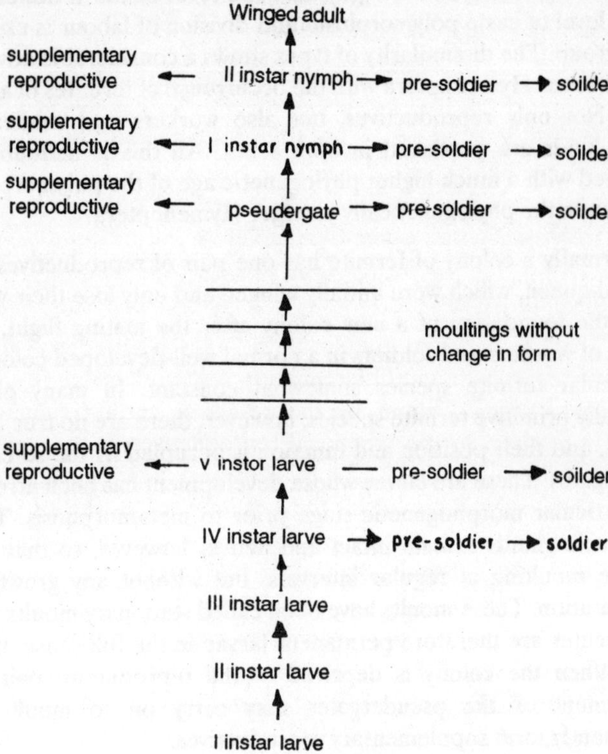

with wing pads. Further, the nymphal instars retain the potential to develop not to the winged adult but to supplementary reproductives or to pre-soldiers. Regressive moults back to the pseudergate are also possible. Table 2 delinicates possiblities of modification.

In phylogenetically advanced termites the situation is more complex, and various authors have professed class that genetic and histogenetic determination is involved. The fact that some termite species form their workers from other species has been stressed as evidence for genetic determination. Whereas both sexes are able to produce both castes in other species, particularly in the more primitive termite, each sex in the more advanced termites seems secondarily to have lost the potential to produce one or other caste. Species regulation in these species is therefore controlled by sex and in turn, doubtlessly manipulated genetically.

The various species of the family Termitidae differences in internal structure between winged and apterous adults which is by now clear in the first instar. In *Tenuirostritermes tenuirostris* (Desneaux) the inhibitory affect of reproductives on the formation of further reproductives has been observed at the time of hatching and according to Weesner, may occur even earlier.

The inhibiting substances

After some years of experimentation it has been proved that inhibition of the growth and differentiation of pseudergates (e.g. in *Kalotermes*) is brought about by the result of certain chemical substances produced by the functional reproductives; the so-called queen inhibitory substance, and probably also a king inhibitory substance, these substances inhibit the development of ovaries (and testes) and doubtless at the same time inhibit growth (including differentiation) of the imaginal parts of the body and of the body as a whole. Therefore, by the exercise of the inhibitory substances the development of the majority of the larvae in a colony is checked at the pseudergate level. When the reproductive pair dies or is artificially removed from the colony the inhibitory effect disappears and many of the pseudergates change to supplementary reprodutives at the next moult. The changeable pseudergates are in a suitable physiological state effect disappears, i.e., at the appropriate stage of the intermoult period. The same applies to larvae in the appropriate stage of the Vth larval instar and to those in the Ist and IInd nymphal instar Just as the first pair of supplementary reproductives starts to reproduce, further supernumerary reproductives are killed by the workers and the inhibitory substances produced by the new king and queen constrain the consequent distinguishing factors in the procreation in the colony.

Luscher's (1961a, b) up-to-date papers suggest the existence of four defferent inhibitory substances concerned with regulating the development of the supplementary reproductives. The effect of at least two of them seems to be amplified by the workers (pseudergates) through a remarkable sex-linked effect; the male pseudergates seem selectively to inhibit the development of female supplementaries and *vice versa*. Supplementary proofs, both probationary & comparative is compulsory before doubtful understanding the conclusions and the effects.

The discovery of the source of the inhibitory substances has not yet been ascertained. Experiments in which a wire mesh prevents pseudergates reaching either the anterior or the posterior parts of the body of living reproductives, and others involving the use of extracts from various parts of the body, suggest that the active substances originate either in the head or thorax of the specimen and that they leave the body via the intestinal tract, most probably with the faces. Both, female and males produce different inhibitory substances. The substances produced by the king and queen are complementary in such a way that the full effect is only obtained if both reproductives are present. Comparativelly, the inhibitory effect of the female is feeble while that of the male is almost non-existent.

Almost nothing is known about the chemistry of the inhibitory substances. The active principle can be extracted from the queens with methanol or with water, and there is little difference in the activity of the two extracts. Even after the death of the animal, the substance remains active for several months, yet the effect of the repro- ductives disappears within 24 hours of their removal from the colony. This has been interpreted as a rapid inactivation of the inhibitory substance although the substance may be quickly removed from the circulatory system by the Malpighian tubules and may perhaps be inactivated during its passage through them. Büchli (1956) has shown that the substance (substances) is only active when applied during the first third of a particular instar but that the state of the fat body (quantity of reserve materials) also seems to be important. It is doubtful whether the substance can also be produced by specimens which are unable to reproduce although some experiments suggest that this is so. Lüscher's (1955) experimental evidence shows that these substances are exohormones. When a sexual pair were separated from pseudergates by a double barrier of wire gauze which prevented any contact but allowed the passage of odours, the result was equivalent to the complete removal of the reproductives: a pair of functional reproductives was produced and they survived. Every evolving additional reproductives were eradicated when only a signle barrier allowing antennal contact was used. It constraining from the experiments whether the substance (or substances) perceived by the antennae is identical with the substance (exohormone) inhibiting the development of reproductives or whether it merely affects the behaviour of the pseudergates and causes them to eliminate the excess reproductives. In the latter case the substance would be a pheromone in the sense used in this book. Bees have the similar coextensive action of exohormone and pehromone.

The stimulating substances

Both, the stimulating and inhibatory substances in winged and supernumerary reproductives, Lüscher (1956b, c) claims to have found one or perhaps two substances which have a stimulating effect on the development of supplementaries. He was able to extract this substance from the heads of supplementaries with methanol, and strips of filter paper soaked with the extract proved to be active. Contrasted with control experiments, the applied extract resulted into a clear cut affirmative effect though only if the reproductives were removed. The experiment shows that the inhibitory effect is stronger than the stimulating effect. Lüscher assumes that one of these different sex-specific substances is produced by the females which stimulates the production of males, and the other substance by males stimulating the production of females. In the natural colony these substances would assume importance if one of the reproductive pair were lost. The presence of the remaining reproductive would then stimulate the replacement of its mate by a supplementary of the appropriate sex. The selection value of the stimulating substances would probably lie in the acceleration of the replacement process and it appears to be all the more significant as the deprivation of only one reproductive is much more commonly in natural conditions than the simultaneous loss of the pair.

Exohormones and the metamorphosis hormones concurrent

Many countries have examined the feasibility that exohormones modify the secretion and activity of the metamorphosis hormones in termites. Kaiser (1955) has shown that discrimination from the pseudergate level in the more primitive termites involves a different number of moults for different castes, i.e. three for a normal winged reproductive, two for a soldier and one for a supplementary reproductive. These variations indicate corresponding differences in the production and action of the metamorphosis hormones, and histological differences were in fact found in the ventral glands and the corpora allata. In the more advanced termites, the evolution of workers is correlated with reduced functioning of the ventral glands while the development of soldiers is correlated with a notable increase in size of the ventral glands and corpora allata. The ventral glands degenerate following the imaginal molt. A large increase in the size of the corpora allata was observed in each instar during the development of supplementaries, but this increase was not found during the development of normal reproductives where the increase occurs at the time of egg-laying. During stationary moults, the ventral glands of the pseudergates were about double the dimension of those in the intermoult period.

Lüscher (1957a) and Lüscher and Springhetti (1960) examined the impact of implanting corpora allata from the other donors into pseudergates in normal colonies and colonies which had lost their reproductive pair. In most cases, the implanation of the corpora allata resulted in moults to pre-soldiers. Under some conditions moults to intercastes between pre-soldiers and supplementary reproductives and between pseudergates and pre-soldiers resulted. The authors concluded that at least three factors influence the type of differentiation which follows implantation of corpora allata. Firstly, the physiological state of the recipient specimen; Secondly, the physiological state of the colony; and Thirdly, the physiological state of the implanted gland (i.e. the physiological state of the donor specimen).

Entomologist advanced the hypothesis that there are two different corpus allatum hormones, one with a juvenilizing effect and the other with gonadotropic activity. According to the theory, 'juvenile hormone' favours the differentiation of supple- mentaries while the 'gonadotropic hormone' favours the differentiation of pre-soldiers. It is rather a difficult theory to expound even in part, the various experimental results obtained by the authors. A simpler theory can be proposed which would comply with GF theory. Such a theory involves the assumption of only one corpus allatum hormone, which does not disagree with anything known about JH. In the presence of the inhibitory substances (i.e. in the presence of the reproductives) growth (both isometric and gradient growth), as well as production of JH, are more or less completely inhibited. Under these conditions a low dose of JH favours development towards supplementaries while increasing doses (prolonged period of action) result in intercastes and eventually the differentiation of pre-soldiers. If the exohormones are not present, supplementaries will develop. This prothetelic effect seems to be absolutely because of a reduced proportion of nutrients for the imaginal parts by the accelerated growth of the ovaries in the absence of the inhibitory factor, in accordance with the law of correlation of consumption. Existence of inhibitory factor the application of JH will again induce the formation of pre-soldiers with possible intercastes at the expense of the supplementaries. Another piece of evidence in favour of the GF theory is that the ventral glands do not degenerate until a definite level of differentiation, as in supplementaries and adults, has been reached. There action, both in pseudergates, where differentiation is inhibited by the exohormones, and in pre-soldiers, where differentiation is delayed by JH. The dissociation in pseudergates of chronically repeated moulting from growth and differentiation is additional confirmation in support of the GF theory assumption that the two processes are independent and that MH acts directly only on the former.

Lüscher (1969) supported by the experiments with Röller's JH preparation, and he discasded the hypothesis that there are two or more different JH, after demonstrating that the same JH can induce both effects. In place of his earlier views, he adopted the new hypothesis that these effects are caused by different doses of the same JH produced and communicated to other members of the colony by individuals belonging to different castes, so that it would thus have the character of a true exohormone. In such a condition consistent only one inhibitory hormone would be necessary, i. e. the one stabilizing the pseudergates and inhibiting soldier development. It has been suggested that is a JH inhibitor, but its effect must be more general, i.e. inhibition of any growth and morphogenesis. This is with the findings of Springhetti (1960) and Lebrun (1967, 1970). For instance, the latter author showed that the implantation of *Periplaneta americana* corpora allata into nymphs developing into adults (Ist and IInd instar nymphs) became the cause of the growth of intercasts amidst sexual adult and soldiers.

The phylogeny of social life in termites

The picture of origin of social life in the Isoptera has been clarified by the findings of exohormones. As illustrated by Kaiser (1955) pointed out, the termites differ from the Hymenoptera in that no examples are known of a solitary way of life or of transitions between this and social life. The first signs of a development towards a social life can, however, often be seen in the orders Dictyoptera (in the suborder Blattodea) and Dermaptera which are closely allied to the ancestors of he termites. In these orders it is a common occurance that nymphs in various instars live together with the female or with the parent generation. When such situation occurs the appearance during phylogeny of inhibitory substance of an exohormone nature would necessarily prolong the situation by keeping the morphogenetic stage of the nymphs static. This would automatically keep the stage of development of the brain unchanged and thus preserve the nymphal instincts. The selection value of each of these effects as a step towards social life is clear. The initial stages would prepare the way for the development of trophallaxis and other features of social life (cf. Schneirla, in Roeder, 1953). The results are confirmed by the wide non-specificity and wide distribution of the exohormones as shown by several authors.

EXOHORMONES IN THE HYMENOPTERA

Polymorphism

Polymorphism in the Hymenoptera significantly differs from that in termites. The essential variation is that (apart from perhaps a few unimportant exceptions) only the female sex is involved. Usually only one caste besides the reproductive females and males is found; i.e. the female workers, which may in some cases be polymorphic. A second sterile female caste, the soldier, is only found in some species of ants. In the colonies of the honey bee (*Apis mellifica*) only one extra caste exists. These are females with reduced ovaries which are known as workers. The first differentiation to worker or queen larvae does not start until the third day after the larva hatches. Till this moment any larva may develop in either direction. The determining factor involved is the quality and quantity of the food supplied by the workers. However, normally the fate of each individual is already decided at the time of egg laying, as the eggs from which the queens develop are laid into special large and distinct queen cells. In the adult, the ovaries of workers in colonies with queens are reduced and functionless, whereas in queenless colonies the ovaries of the workers become functional and the workers start laying eggs. It is now established that both ovarian development in the workers and the consequent instinct of workers of construct queen cells are controlled by definite substances produced by the queen which are similar to the exohormones and pheromones already discussed in termites. Now it is confirmed the more primitive termites, however, both the bee and ant workers are adult insects in which the prothoracic glands have disappeared and are therefore unable to moult and develop further. However, there is a crucial difference between the bees and ants; whereas the queen bee is slightly neotenic in comparison with the worker bees, while the wingless ant workers are strongly neotenic the winged sexual females are not.

Contemporarly studies by the researchers in this field have elucidated the problem still further, although some questions still require investigation. A complex determinant system is now assumed to control caste determination in honey bee females. The difference in the larval food from the very first day of larval life is the major environmental difference. The food prepared for the larvae in he worker cells (worker jelly) differs from the brood food in the queen cells; in addition to pharyngeal gland secretion, it also contains an inhibitory substance from the mandibular glands of the nurse bees. After about three and a half days, the worker jelly is modified by the addition of honey and pollen. It unmodified jelly is fed to older larvae, it inhibits morphogenesis. The first differences between worker and queen larvae (in the structure of the fat body cells) emerge within the first three days of the larval life.

Till now many differences between the structure and activity of the larval worker and queen endocrine systems have been described. In queen larvae, the corpora allata seem to attain a higher level of activity, and at an earlier stage of development, than in worker larvae. Royal jelly probably activates the endocrine system in some way. The topical application of JHa to 3-day-old worker larvae in worker cells produces a number of queen characters in the treated specimens. The effect of queen jelly on the determination of classifications thus seems to be mediated by modified JH production. A worker-like appearance of the fat body cells can be induced in larvae reared in *vitro* by the addition of mandibular gland secretion.

The inhibitory substance of the honey bee

The fertilized queen produces the inhibitory substance. The first effect of removing the queen from the colony is that the workers construct emergency queen cells. These queen cells are constructed above some of the original worker cells in which eggs or young larvae are still present. If eggs or larvae, both are not obtainable to the workers in the colony when the queen is lost, the ovaries of many (of about 10 per cent of the workers within a week of removing the queen) start developing. After some time the workers begin to lay eggs, but of course these remain unfertilized and only drones develop. Both the construction of queen cells and the development of ovaries in workers are the symptoms of absence of the inhibitory substance, i. e. the absence of the queen. There is also a definite positive effect of the substance (or group of substances) which causes the queens (alive or dead) or extracts prepared from queens to be highly attractive to the workers. It is debatable whether Butler's original assumption (1954) is correct that only one queen-inhibiting substance produces all the effects mentioned, or whether there are at least two substances are suggested by Pain (1952, 1954a, b), Groot and Voogd (1954) and Voogd (1955), one substance being the inhibitor of ovarian development and the other Butler's 'Queen Substance', which Brian (1957) has called a hormone and a signal.

The second view is more acceptable after the latest discoveries which includes Butler's experimentaion. On the other hand, several pheromone- type substances interfering with various kinds of social behaviour in the honey bee have been described. Now it is indubitably proved that the mandibular glands in the queen is the origin of the inhibitory substance.

Pain (1961) examined the three aspects of the effects of the queen on the workers, and suggested that the inhibitory substance is complex and amalgamation of many acids.

Pheromone I. The substance isolated by Barbier (1958) from active extracts of queens and workers in colonies with queens; a substance also isolated by Callow and Johnston (1959) from the heads of fertilized queens. Both groups of authors simultaneously synthesized this acid: Barbier (1960) synthesized it from cycloheptanone and Johnston (1959) from azelaic acids. It has the following formula:

$$CH_3-CO-(CH_2)_5-CH=CH-COOH$$

Pheromone I

and is α, β unsaturated 9-oxydec-2-enoic acid. Pain called this acid phromone I. The acid forms white flakes and has the following effects on queenless workers:

(a) It constrains the construction of queen cells – at a concentration of 0.13 μg per bee according to Butler and at 0.5 μg according to Pain (1961a).

(b) It is a crucial component of the queen odour but it is not itself attractive to the workers.

(c) It does not stop the formation of eggs in the ovaries of workers when injected, or in food or when licked from the body surface of the dead queen.

Pheromone II. Secondly, the substance which Pain (1961) called pheromone II was isolated in the form of highly volatile esters by distillation from the active pentane soluble fraction (pentane acids). Couple of these esters were identified by gas-chromatography as methyl phenylacetate and methyl phenylpropionate. This substance is not attractive to workers by itself, but when mixed with 'pheromone I' gives the characteristic queen odour.

The element which was afterwards recognized as trans-9-hydroxydecenoic acid, with the following formula (Buttler and Callow, 1968):

$$CH_3-CH(OH)-(CH_2)_5-CH=CH-COOH$$

Phermone II

It is also produced by the mandibular glands and development of the worker's ovaries is inhibited only in the presence of the two earlier mentioned substances.

The two substances called pheromone I and II were found to occur in the mandibular glands of fertile queens. Only traces of exohormone I were present in emerging queens, and phermone II appeared later. This complied with the lack of attractivity of freshly emerged queens. The rate at which the pheromone complex (I and II) was produced seems to be determined by not only the age of the queen, but also the number of workers accompanying her, i.e. the state of nutrition of the queen seems to be the important factor.

Constructing her reasoning on an analysis of Müssbichler's (1952) work, Pain went on to claim that the substance causing inhibition in ovarial development of the workers was not identical with the inhibitor of queen cell construction. She was absolutely right in attaching importance to the effect of what she calls nutritional castration of the workers by the queen which causes the inhibition of ovarian development. The researches done previously have shown that the queen influences the development of ovaries in the workers, first by herself consuming a quantity of the pharyngeal gland secretion (royal jelly) of her attendant bees, and then secondarily, and what is much more important, by producing larvae which are the main consumers of royal and/or worker jelly. This elementary mechanism, which works alone with the inhibitory substance, is quantitatively the primary reason of suppression of ovarian development in workers.

However the theory which is less convincing is Pain's conclusions (1961) on the olfactory character of the queen inhibitory substance, which she suggested worked by inhibiting the secretion of the corpora allata, but her findings do not negate the existence of a special inhibitory substance of the exohormone type, which directly and biochemically suppresses the development of the ovaries. Such a substance has been suggested in the papers of Butler (1954, 1957), Carlisle and Butler (1956) and Hrdy, Novak and Skrobal (1961). Thus the honey bee queen probably produces at least two pheromones (I and II of Pain) and one exohormone. Therefore it seems logical to accept that the inhibitory effects are caused by the three substances mentioned above.

Besides the views expressed by Pain (1961) other authors seem entirely to agree that actual contact is necessary for the substance inhibiting ovarian development to be transmitted. This illustrated by the fact that a double wire screen between the queen the workers prevent this inhibition. That the effect of the substance is weaker when taken with food than when licked from the body surface of a dead queen is in no way contradictory of the purely direct biochemical action that Butler (1957, cf. Karlson and Butenandt, 1959) suggested. Both the amount taken per specimen as well as the concentration of the substance seem to be important in producing the inhibition, and both these were probably lower in the food than on the body surface. Thus the grounds for Voogd's psytchic stimulations hypothesis (1955, 1965) is not acceptable.

In opposition to Hinton's conclusions (1955b), the worker bees seem piay a cruial part in the transmission of the exohormone and the pheromone. It has been demonstrated if the queen substance, in the form of an aqueous solution of an alcohol extract, is given to worker bees, ovarian development and queen cell construction are suppressed not only in these workers, but also in others which had not initially conscemed the extract.

The method of exercise of the *queen inhibitory substance* on the worker bees was studied by Lüscher and Walker (1963). They found that in workers separated from the queen the corpora allata increased in volume. Such proliferation was significantly inhibited by the queen extract prepared from about 100 queen bees by Butler's method. The experimenters concluded that the inhibitory effect of the queen substance on the ovaries of workers was the result of the suppression of the secretory activity of their corpora allata.

The new discovery concerning the honey bee is reported by Butler, Callow, and Chapman (1964). Theydemonstrated that a synthetic substance, 9-hydroxydec-trans-2-enoic acid, caused dispersing clusters of swarming honey bees to re-form and settle down again in similary as does the odour of the queen, i. e. her mandibular gland secretion.

The occurrence of exohormones in other Hymenoptera

There is almost no information on exohormones in Hymenoptera other than honey bees. According to Sakagani (1954) the effect of the queen substance (or substances) in the Japanese bee *Apis indica japonica* seems to be much weaker than in *Apis mellifica*, for fully developed ovaries were found in seventeen out of twenty workers accompanying the queen. This probably the effect of the workers taking some of the royal jelly, i. e. the secretion of the pharyngeal glands which is fed to the queen. In this species the stimulating effect of the jelly appears to be stronger in comparison to the inhibitory effect of the queen substance.

The observations made by Cumber in 1949 suggests that substances with an inhibitory effect on-ovarian development also occur in bumble bees. He found that the sterility of auxiliary females was not decided by the quantity of food available to the bumble-bee colony but by the presence of the queen, for removal of the latter resulted in development of the ovaries of the auxiliary females. Similarly the loss of the queen in colonies of *Polistes gallicus* resulted in egg-laying by the 'workers' within about 24 hours corresponding effects have been found in some species of the genus *Halictus* (Que, 1958)*.

Another discovery was made by Stumper in 1956 has found an exohormone system, similar to that in bees, in a number of ant species. This system seams to control the following behaviour pattern: (1) crowding of the queen by the workers; (2) extensive licking of the queen; and (3) carrying the queen.

Resembling out come are caused by the same substance in various Hymenoptra, and are especially pronounced in *Lasius alienus* and *Pheidole pallidula*. Petroleum ether proved the most suitable solvent for extracting this exohormone. A dead *Lasius* worker impregnated with an extract of a *Pheidole* queen was accepted as a queen by an orphan *Pheidole* colony. Bier (1956, 1958) has shown that the presence of functional queen inhibits ovarian development in workers of the species *Leptothorax unifasciatus* and *Formica pratensis*.

Those female wasps and bumble bees which do not lay eggs are not true workers as they are in honey bees and ants. They are merely unfertilized females of reduced size and woth undeveloped ovaries, and are morphologically identical with the 'queen' in every respect, including the activity of gonads.

Metamorphosis hormones cause the classification as suggested by Brian (1958a) in the ant *Myrmica rubra*. He has shown that there is a critical period in the last larval instar of all female larvae during which the functioning of the corpora allata may be prevented by the neurohormone (AH?) of the neurosecretory cells of brain. After such occurance only sterile workers develop. In the absence of this inhibition the increase in secretory activity of the corpora allata promotes the growth of the imaginal wing and ovarial discs and the ovaries, so that sexual females (queens) develop. Development, however, is also dependent on the state of nutrition. Starvation causes a more or less limited development of the wing discs at a critical periods, resulting in intercastes of various grades.

In 1961 Brian later suggested that when the MH (*ecdysone*) concentration increases quickly the relatively small ovaries and wing discs do not have enough time to respond, and workers develop. On the other hand, when the MH concentration increases gradually, as in diapausing larvae, the corpora allata are active and the ovaries and the imaginal discs of the wings grow and develop normally so that winged sexual-queens are produced. The determination of diapause which consequently influences the type of neurosecretory activity of the brain cells thus also affects the classification of the female.

Brian (1959) another thing discussed by the possible existence of a special substance with the character of an exohormone which would simul- taneously suppress the development of ovaries and other female features and influence the instincts of the workers. He concluded that it would also cause the suppression of the secretory activity of the larval corpora allata simultaneously.

The idea was developed further which was influceced by the further discoveris made by Brian on the effect of starvation on queen determination and the effects of the presence of a fertile queen on worker development. It was found that the *Myrmica* queen produced (also in the mandibular gland) a substance related to, but not identical with, the honey bee queen inhibitory substance (Brian and Hibble, 1963) and that 9-oxodec-trans-2-enoic acid was likewise partly effective in *Myrmica*.

The presence of the two effective principles in the mandibular glands of the *Myrmica rubra* queen was suggested by Brian and Hibble (1963). One, probably 9-oxydec-enoic acid, is supposed to produce a stimulatory effect on brood growth similar to that caused by the presence of the queen, the other, which probably corresponds to a lipoid fraction of queen mandibular gland extract, has an inhibitory effect. In another paper, the authors analyzed the influence of the queen on the growth of the brood in connection with the size of the larvae in Myrmica (Brian and Hibble, 1963b). In 1963 Brian found that in the classification of female larvae, the factors such as tempesation, food and the presence of queens is highly important.

CONCERNING THE PHYLOGENETICAL ORIGIN OF EXOHORMONES

The hormonal substance which inhibit the ovarian development and other growth processes in arthropods is not comfined to the social insects. It has been known for a long time that neurohormones produced by the eye stalks of many Decapoda have an inhibitory effect on moulting and growth Therefore, it is interesting to note that Carlisle and Butler found that the ovary-inhibiting hormone in the eye stalks of Decapoda produced an effect similar to that of the queen substance on the ovaries of honey bee workers in the absence of the queen. The consequence is equivalent to that obtained on ant and termite workers. In same manner, the queen substance has been shown to suppress moulting and the functioning of the ovaries in prawns. It has been assumed that the sunstance mentioned above are steroids (Carlisle and Butler, 1956).

The already accepted uncertainty of the queen-bee inhibitoy substance has been further confirmed by Hrdy, Novak and Skrobal (1961). An aqueous solution of an evaporated alcohol extract of fertile honey bee queens was tested on groups of *Kalotermes flaviocollis* workers. The considerable delay cauesd by the extract the development of supplementaries compared with control groups.

3

The Neurohormones

The term, Neurohormones is largely said to comprise all the physiologically active products of the nerve cells. Not with standing the quantity of its secretion, each nerve cell secretes actively in this sense, at least at its nerve ending, synapses, etc. The enzymatic synthesis and decomposition of specific materials such as acetylcholine and nor- adrenaline at both sensory and motor nerve endings and synapses is the source of changes in the electric potential. Nervous activity is based on there change these substances pro- duced by normal nerve cells were included in the term neurohormones. However, in the recent years the term has been given a new and specific meaning to omit the aforementioned ultramicroscopic secret- ions, for which the term neurohumoral factors is used. The neurohumoral factors are not true hormones for the purposes of this book as are effective at their place of origin and thus belong to the category of protohormones. The neurohormones *sensu stricto*, on the other hand, are a special group of true hormones combining the characters of both tissue hormones, being produced in tissues with functions other than hormonal, and gland hormones, originating in specialized secretory (neurosecretory) cells. From the true neurosecretory cells there can further be differentiated by the cells of nervous origin which have become completely specialized to endocrine neurosecretory activity and have lost their nervous function during this phylogenetic adaptation, e.g. the cells of the medullary part of the adrenal bodies producing noradrenaline. There is, however, a clear phylogenetic link between these three types of cells which is also clear from their evolution in ontogeny. It proceeding from phylogenetic considerations that the same active substance may occur in one case as a neurohumoral factor in the endings of normal nerve cells, and in another as a neurohormone in a neurosecretory cell.

The definition of the actual neurosecretory cell can be stated as as a special type of nerve cell characterized, in addition to its nervous function, by the production of a secretion with specific properties. This secretion originates in the cytoplasm of the neurosecretory neurones in the form of granules of variable size, characterized by their reaction to particular staining methods and by their shining white appearance under dark-ground illumination. Another signifiacant aspect of the neurosecretory granules observed in most of the known neuro- secretory cells is their movement along the axons of the neuro- secretory cells, usually to places where they accumulate, i.e. the paired neurohaemal organs, e. g. the corpora cardiaca of insects, the neurohypophysis of vertebrates or the sinus glands of crustaceans. From the site of àccumulation they pass into the circulatory system at specific periods. This movement of neurosecretion has been shown experimentally in the neurosecretory cells of various animals, includ- ing insects. As stated earlier the cutting of one of the vervi corporum cardiacarum results in the accumulation of the neurosecretory materials in the cut end of the proximal part of the nerves, whereas it gradually disappears in the distal part including the corpus cardiacum and corpus allatum. The movement of granules has also been observed in living cells in tissue cultures (Hild, 1954). Observation shows that the speed of movement is about 3 mm in 24 hours.

Another method has been illustrated by Johnson (1963) in which a neuro- secretion can reach its target tissue. Based on the of observations in aphids, he concluded that the neurosecretory substances originating from the brain neurosecretory cells pass through the corpora cardiaca and are distributed to various internal organs inside the nervous pathways leading from these structures. This result is based on convincing histological evidence obtained in several aphid species, in which the granules could be observed along the branches of all three nerves leaving the corpus cardiacum, right to their endings in the muscles and other organs. The attempt made by Johnson to generalize his findings and suggested that the reason why similar observations in other insects bore negative results might be that 'in insects other than aphids, either insufficient material is normally present in the axons for it to be readily detected by histochemical means, or it does not stain differentially with the known straining procedures'.

On the other hand the crucial evidence contradict this conclusion. Firstly, there is irrefutable experimental evidence of the role of the haemolymph in the action of some of the brain hormones. Secondly, organs such as the corpora allata, supposed by some authors to be regulated by the import of neurosecretory material, show the same cyclical activity after transplantation, indicating that activation hormone is present in the haemolymph. Thirdly, there is even less evidence of the direct utilization of neurosecretory material in tissue than of its passage into the haemolymph. Finally, the demonstration made by Ganguly and Banerjee (1961) in the pyrrhocorid bug *Macrocercea grandis* should be stated here.

The mone probale suggestion is made by Highnam (1965, personal communication), that the condition found in aphids is an exception among insects, perhaps because far-reaching changes in the amount of body fluid associated with sucking plant sap, which could otherwise unfavourably influence the hormonal balance. It can be deduced that the development of neurohaemal organs such as the corpora cardiaca is a later acquisition during evolution, so that the state in aphids could simply have originated as a return to a more primitive stage as is the case for a number of other aphid structures. But it may not certainly mean that direct transport through nerve endings does not also exist, in many other groups of insects, in addition to humoral transport, since it is unlikely that neurohaemal organs like the corpora cardiaca in aphids should be without any function.

The first suggestion was made by Ernst Scharrer (1952), the sum total of the neurosecretory cells in the organism, together with their specific neurohormones, can be regarded as a special, third correlation system of the animal organism mediating relationships between the nervous and endotrine systems. Its function consists in mediation between slow, repeated, prolonged environmental stimuli of low intensity, e.g. changes in photoperiod, and long-term reactions by the transformiation of similar, weak nerve impulses to slow endocrine responses resulting in physiological adaptation of long duration, e.g. gonadal activity. The contemporary research is conuntrated primarily on various transport mechanisms and release mecha- nisms at the ultra-structural level indicating the need for a new classification of modes of communication between neurons and non- neuronal effectors.

The carrier substance

Not much information is available concerning the chemical composition of the neurohormonal granules. But there is ample evidence to prove that the granules are not pure neurohormone; they are composed of stainable material, the carrier substance, on which the corresponding neurohormone is bound either chemically or physically e.g. adsorbed. The neurohormones seem first to be liberated from the carrier substance when passing into the blood (haemolymph), perhaps freed by the enzymo- lytic activity of the haemocytes. Many forms of neurosecretory cells can be distinguished by the staining properties of he carrier substance. No data are available on the connection between these differences in the carrier substance and differences between the neurohormones in the different neurosecretory cells. In contrast with its original concept, the carrier substance is now supposed to be insoluble in lipoid solvents and is probably polypeptide or protein in character. The molecules of the neurohormones, relatively small in size, seem to be attached to the large protein molecules of the carrier substance. The neurohormones themselves seem to have rather different chemical properties. Some of them have been said to have lipoid character, e.g. a substance with the effect of AH was claimed to be identical with cholesterol. This has since been questioned by Gersch (1961) by his results on the identify of AH with the neurohormone D_1. Other neurohormones have been shown to be substances soluble in water and alcohol with acid or alkaline proper- ties (e. g. neurohormones C and D,). Leaving the impractical theories the opinion that at least some insect and other invertebrate neurohormones, like the main vertebrate hormones, are of a polypeptide nature now seems to be thoroughly substantiated. Indirect evidence of its veracity was obtained in the eyestalks of crustaceans. The suggestion made by Novak and Gutmann (1962) indicate that the carrier substance is closely related with the so-called gliosecretion (gliosomata of B.) of *Periplaneta* and other arthropods. This independent phylogenetic origin of the carrier substance and the neurohormones indicate that perhaps the carrier substance represents the original secretory activity of the primary ectoderm cells.

A comprehensivecytochemical study of the neurosecretory granules and other neuroplasmic inclusions in *Periplanta americana* was carried out by Pipa (1961, 1962). In this, gliosecretion, as well as other structures, were discussed. In another paper the ultrastructure of the glio- secretory granules was investigated. The differences found between these and the neurosecretory material seem to be irrelevant from the point of view of the conclusions reached by Novak and Gutmann (1962), as suggested by Novak (1964).

The occurrence of neurohormones

Histologically typical neurosecretory cells have been found in all the main metazoan Phyla. In insects, they have been found in all groups where they have been seriously looked for. This also applies in crustaceans and either arthropods. Now it is certain that at least three or four different neurohormones occur in all arthropods.

The general significance of neurosecretion in the organism

Neurohormones act either directly on the effector organs, as in myotropic or chromatophorotropic effects, or indirectly by the secretion of secondary hormone (e.g. the MH), or again by influencing the production of a territory hormone regulated by the secondary, dependent hormone. The latter applies to the control of the adrenal hormone in vertebrates by the ACTH of the adenohypophysis which is itself under the control of the hypothalamus (cf. Scharrer, B., 1959). Many instance of environmental influences mediated in such a way have been observed in insects, e.g. the influence of the distinction of the abdomen on AH production in *Rhodnius* and the corresponding influence of photoperiod in diapausing insects, etc. Undoubtledly the various cyclic phenomena in the production of myotropic and chromatophorotropic effects observed in various insects by Gersch *et al*. (1960) belong to this particular classification.

The acceped theory of the photo-neuro-endocrine systems and the question of their evolution is discussed by E. Scharrer (1964). He emphasizes the role of light as a synchronizer of circadian and annual rhythms in many important functions of living organisms, such as reproduction, moulting, migration, metabolic adjustment, etc. He differentiasted the various steps in the evolution of complex neuroneurosecretory-endocrine control, as known in vertebrates: (i) the transparent animal in which the organs are exposed to changes in illumination; (ii) animals with an opaque integument excluding light except through a transparent window overlying the central nervous system (in insects, for example, this is known in the pupa of *Antherea pernyi*); (iii) animals in which the retinal and central neurones transmit stimuli directly to endocrines affecting the light-dependent target organs; (iv) animals in which the products of the neurosecretory cells intervene between the nervous and endocrine systems. However,he insisted, that photoperiodicity is only one of the phenomena handled by neuroendocrine mechanisms.

Lot of research work is associated with the ***general signi- ficance and origin of the neurosecretory cells***. Beln (1963) views the 'neurosecretory neurone' as a doubly specialized cell having the features necessary both for the conduction of impulses and for the production of secretion. He discusses the ultrastructural characteristics of neurosecretory cells, with special reference to the formation of elementary neurosecretory granules by the Golgi apparatus and to the fate of neurosecretory granules in nerve endings (the so-called synaptic vesicles of the neurosecretory cells). He futher states that since the membranes of the neurosecretory axon bulbs are consistently smooth-surfaced, there is no evidence of pores or canals which might serve as channels for the discharge of neurosecretory granules formed, and no definite evidence of the extraneuronal occurrence of neurosecretory granules. But unlike Johnson he puts more stress on the hormogenic function (in the sense of a true endocrine organ) of the neurosecretory cells as one of their unique features. He also considers the possibility of the factors which may decide the development of specialized 'hormone-producing neurones amid their ordinary neighbours'.

The effects of neurohormones

If we reject theexistence of the neurohumoral factors concerned with the nature of nervous activity, the following categories of effects caused by neuro- hormones can be differentiated: (1) Myotropic effects (influencing the frequency and amplitude of the spontaneous rhythmic activity of muscle, e.g. in the heart, Malpighian tubules, oviducts, gut, etc.). (2) Chromatophorotropic effects (influencing pigment movements, i. e. dispersion and concentration, in epidermal cells and chromato- phores; also the development of various pigments in the epidermis). (3) Daily activity pattern. (4) Influencing water balance. The activation of enzymes (midgut proteinase activity). (6) The activation of the secretion of other endocrine glands. (7) The induction of diapause (the humoral effect of the suboesophageal ganglion in the female moth pupa on the diapause of the embryos of the next generation. (8) Influencing the development of the gonads and related effects, e. g. the suboesophageal castration cells in females of *Leuco- phaea maderae*. Surely, it is logical to conclude that several of these activities are due to one and the same hormone. Thus, for example, the myotropic neurohormones C and D of Gersch have been shown to produce corresponding effects on pigment movements in *Carausius* and *Corethra*. Possibly even if they are caused by the same or different hormonesone of these hormones (D_1) is identical with the AH of the pairs intercerebralis. But as there is till little clear evidence regarding the other activities it seems preferable to discuss the various activities separately.

Contrary to the metamorphosis hormones, insect neurohormones came from various groups of neurosecretory cells (nsc) detected by histological methods, but the functions of the relevant neurohormones are not known. On the other hand, the effects of extracts of various parts of the nervous system are known, although the nsc which produce them have not been identified. There are very few exceptions to this rule, the chief ones being the activation hormone produced by nsc of the pairs intercerebralis and the diapause hormone produced by the paired groups of cells in the suboeso- phageal ganglion. Considering the aim of this book it would be suitable to include the distribution and various features of the neurosecretory cells and the corresponding structures in connection with their nerve endings which must be discussed separately from individual functions whose origin is not yet discovered.

DISTRIBUTION OF NEUROSECRETORY CELLS

The Cerebral Ganglion

The most important paired group of nsc is localized within the pars intercerebralis. Normally It contains several types of nsc, as described by Johansson. Apart from the obvious ones, which are generally held to be the source of AH, there are a number of smaller, less conspicuous cells to which various other neurohormones or functions have been ascribed (e.g. Girardie). The structure, staining properties and ultrastructure of various nsc were studied in detail by Panov and his co-workers in different insect groups and Raabe and her co-workers. Gersch has carried out an immunological study of the neurosecretory system of Periplaneta americana. He discovered high specificity against the neuro- secretory material and the antibodies selectively labelled the secretory neurones in the brain and the secretory products of its nsc in the retrocerebral complex.

The comprehensive account of the neurosecretory system in the head of pyrrhocorid bug *Macrocerea grandis* (Gray) was given by Ganguly and Banerjee (1961). In addition to the neurosecretory cells already known, three new groups were described and their homology with the X-organ, sinus gland and epineurium of crustaceans was claimed. The term *Wigglesworth organ* was suggested for the supposed X-organ homologue and the name *Hanstrom organ* for the hypothetical sinus gland homologue. Both structures were later rediscovered in the brain of the adult silkworm (*Bombyx mori*) by Ganguly and Basu (1962a). A elaborate cytochemical analysis of the various types of neurosecretory cells was carried out in the last mentioned and the research papers which were published later (Ganguly and Deb, 1960; Ganguly and Basu, 1962b).

The formation of the neurosecretory system and its association with diapause in the last larval instar were studied in the moth *Ostrinia* (= *Pyrausta) nubilalis*. Uncertain variations were discovered between the amounts of neurosecretory material in diapausing and nondiapausing brains in any of the groups of cells. A comparative cytological investigation of the brain neuro-secretory cells was made by Panov and Kind (1963) in 20 different species of Lepidopytera. The cells which are divided into six group were found in each hemisphere of the suboesophageal ganglion and the similarities in their structure was found in all the species studied.

A succession of exhaustive investigations on the brain nsc in the water-beetle *Dytiscus marginalis* and their daily and seasonal cycles of activity were published by Abraham (1964, 1966).

Entomologistd have studied cyclical changes in these cells in different insects during larval instars and metamorphosis, at both the optic and the electron microscope level. Accumulation of neurosecretory granules, which differed in different types of neurosecretory cells, was observed during diapause. The author concludes that some of these cells have a specific role in diapause development and that diapause is not a simple consequence of the absence of AH (called 'growth and development hormone' by the author). But there is difference must be made between accumulation granules in the neurosecretory cells and actual secretory activity, including the axoplasmic movement the materials produced and their passage into the haemolymph. As shown by Highnam (1965), the actual rate of the endocrine activity of a neurosecretory cell is affected by the following three factors, which may vary independently: (i) the rate of synthesis of material in the cell; (ii) the rate of transport of the material along the axon; (iii) the rate of release of the transported material into the circulatory system. Thus, we find that interaction among them is difficult.

Four particularly large neurosecretory cells have been described by Girardie and Girardie (1970, 1972) from the protocerebrum of *Locusta migratoria*, located in the anteroventral area of the protocerebrum, beneath the nerve of the median ocellus. Their axons join the nervi corporum cardiacarum I. They are positive to paraldehyde fuchsin, alcyan blue and Victoria blue. Their electrocoagulation results in the growing volume of the haemolymph and with a distension of the abdomen. The implantation of a set of these cells compensates these changes. The innervation of the lateral ocelli and their connection with the retrocerebral neuroendocrine system by means of the nervi corporum cardiacarum II was studied by Brousse-Gaury (1971) in the same laboratory.

The suboesophageal gang

Many paired groups of neurosecretory cells to have been described in the suboesophageal ganglion in various insect orders. Among the nsc in the suboesophageal ganglion of female cockroaches (*Leucophaea maderae*), B. Scharrer (1955a, b) found four type A-cells with a produsion of neurosecretory granules staining purple with paraldehyde fuschin, two ventromedial type B-cells with red-staining granules occurring only on the surface layer of the cytoplasm, and two smaller type C-cells, also with red-staining granules. Fukuda and Takeuchi (1967a, b) described another type of cell in this ganglion, which supposedly produced the diapause hormone: a pair of conspicuous nsc staining red with azocarmine and situated posteriorly to the centre of the ganglion. There is no information concerning the way in which the neurosecretory substance of these cells reaches the target organs, i. e. the ovaries. A distinct difference in their appearance was found in pupae conditioned to lay non-diapause eggs and those laying diapause eggs. In the former they are filled with neurosecretory material, which evidently remains unreleased, while in specimens producing diapause eggs the amount of neurosecretory material is very small, so that it is probably released immediately into the haemolymph. The whereabouts of such occurance has not yet been asurtained

The ventral nerve cord

Information on the outcome and the operation of the neurosecretory cells in the ventral ganglia are still very scarce. The secretory products from all three types of neurosecretory cells found within the thoracic and abdominal ganglia of the roach *Blaberus craniifer* were observed inside axons in the connectives between ganglia. Ligation of the connectives between ganglia showed that the neurosecretion moved in both anterior and posterior directions from the ganglia in which it was produced. A compnehsive research on neurosecretory cells in the ventral ganglia of *Schistocerca gregaria* Forsk, was carried out by Delphin (1963). He described the location of three types of A-cells (A_1, A_2, A_3), two types of B-cells (B_1, B_2), and one type each of C- and D-cells. The axonal transport of neurosecretory material was studied by ligation experiments, in nerve sections and by autoradiographic investigations using ^{35}S- DL-cystine. The experiment illustrated that the granules were passing backwards and forwards along the connectives and along at least some of the nerves originating from the ganglia. Differences in the secretory activity of various cells were observed. Extirpation and reimplantation of the last abdominal ganglion suggested that a hormonal factor which helps to control egg maturation is produced. Egg-laying was notably stimulated by implanting the ganglion from a mature female into a newly moulted one. The author also states that the A_1 neurosecretory cells of the abdominal ganglia secrete an antidiuretic factor which is released under conditions which is necessary for the insect, is that of the water consersvation.

Raabe (1965, 1966) and some of the co-workers another researcher give a detailed description of the nsc in the ventral nerve cord ganglia. Raabe distinguished four categories of neuro-secretory-granule-containing cells in the individual ganglia and paid special attention to the C nsc. These differ from the type stet, B_1 and B_2 nsc by being negative to Gomori's chrome-haematoxylin-phloxine and staining conspicuously only with azan. The distribution of the various types of nsc in the ventral nerve cord of the phasmids *Clitumnus extradentatus* and *Carausius (= Dixippus) morosus* was described in a detail. Here there are 3 pairs of A-cells in the suboesophageal ganglion, 2 pairs in the thoracic ganglia and 1 pair in the first four abdominal ganglia. From the changes observed in their granule content, it was concluded that they have a bearing on the regulation of water metabolism. B_1 cells were found only in the suboesophageal ganglion (1 pair). Three pairs of small B_2-cells were found in each of the abdominal ganglia. It is assumed that their activity is associated to sexual function in females.

The azan-staining C-cells are the most abundant nsc in the abdominal ganglia. They are present in all the ganglia of the ventral nerve cord and also in the tritocerebrum. Two lateral groups of 6 to 8 small cells were found in each of the thoracic ganglia and 20 cells in the central area of the abdominal ganglia. Additionally, some small anterior and posterior cells and a pair of larger posterior cells were reported in the sub- oesophageal ganglion of *Clitumnus*. In this association, special paired neurohaemal organs filled with particles of azan- staining neurosecretory material were found in the paired lateral branches of each ganglion of the ventral nerve cord.

The visceral (sympathetic) nervous system

The widely accepted the insect visceral nervous system stars with the frontal ganglion (in front of the brain) and proceeds posteriorly via the recurrent nerve, to constitute the hypocerebral ganglion, to which the paired corpora cardiaca are joined, being of nervous origin (except their glandular parts, which are of different origin), usually by means of short nerves. At this juncture, one to three pairs of nerves

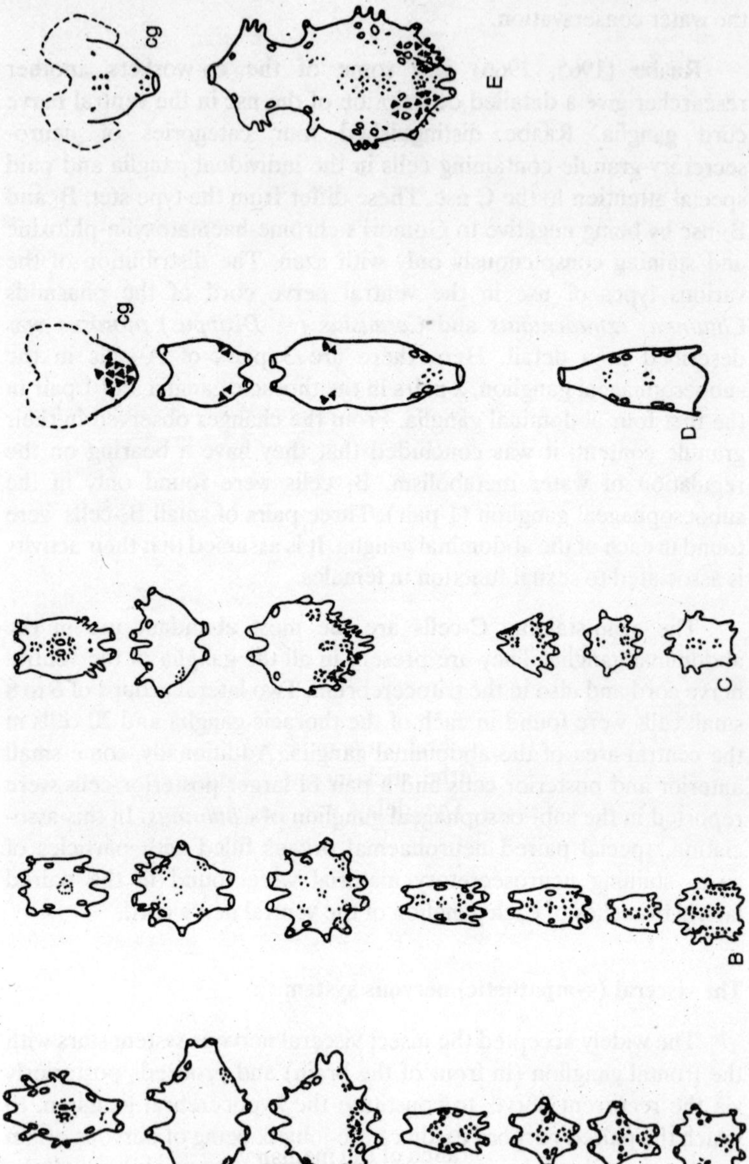
cg
E
cg
D
C
B
A

lead posteriorly to one or two pairs of further small ganglia (visceral and sometimes ingluvial ganglia). The median visceral nerve also originates from the hypocerebral ganglion, conti- nues into the prothoracic ganglion and then connects the individual thoracic and abdominal ganglia in turn.

The information acquired in different groups of insects differ as to the occurrence of the nsc in the individual visceral ganglia. Firstly, it may be because of to the staining methods used, and also to the cycles of their activity, in which there may be periods when the cell perikarya do not contain a significant amount of neurosecretion. Clarke and Langley (1960), using paraldehyde fuchsin and chrome-haematoxylin- phloxine, found no intrinsic nsc in the frontal or any other ganglion of the visceral (sympathetic) system in locusts. van Loof (1960) found nsc in the frontal ganglion of *Bombyx mori*. However, Strong (1966) reported neurosecretory granules in brain nsc axons leading to the hypocerebral ganglion and continuing through it as the inner and outer oesophageal nerve supplying the wall of the forget and the ingluvial ganglia.

The metameric (parisympathetic) neurohaemal organs

Contrasted with crustaceans, no neurohaemal organs except the corpora cardiaca were known in insects until 1965, when Raabe described a pair of specific thickenings in the paired transverse branches of the median visceral nerve in the thoracic and each of the abdominal metameres of the stick-insects *Carausius morosus* and *Clitumnus extradentatus*. The delay was perhaps caused by their inconspicuousness and the fact that they do not stain with specific neurosecretory stains such as chromehaematoxylin-phyloxine and paraldehyde fuschsin. Since then, however, they have been found by Raabe and her many co-workers in most insect orders and found whenever a serious reseach was done.

Morphology

The earlist and simplest case depicts that the metameric neuro-haemal organs (mno) are a pair of symmetrical ovoid thickenings, one in each lateral transverse branch of the median visceral nerve, in front of the ganglion, in each thoracic and abdominal segment. This is the situation in Phasmida (*Carausius, Clitumnus*, etc.), and in some

Orthoptera, e.g. *Acheta* and *Gryllotalpa*. The transverse nerves innervate the spiracles and some of the other pleural muscles. In other cases the thickenings may move closer together and unite at the spot where the branches leave the median nerve, as in cockroaches (e.g.

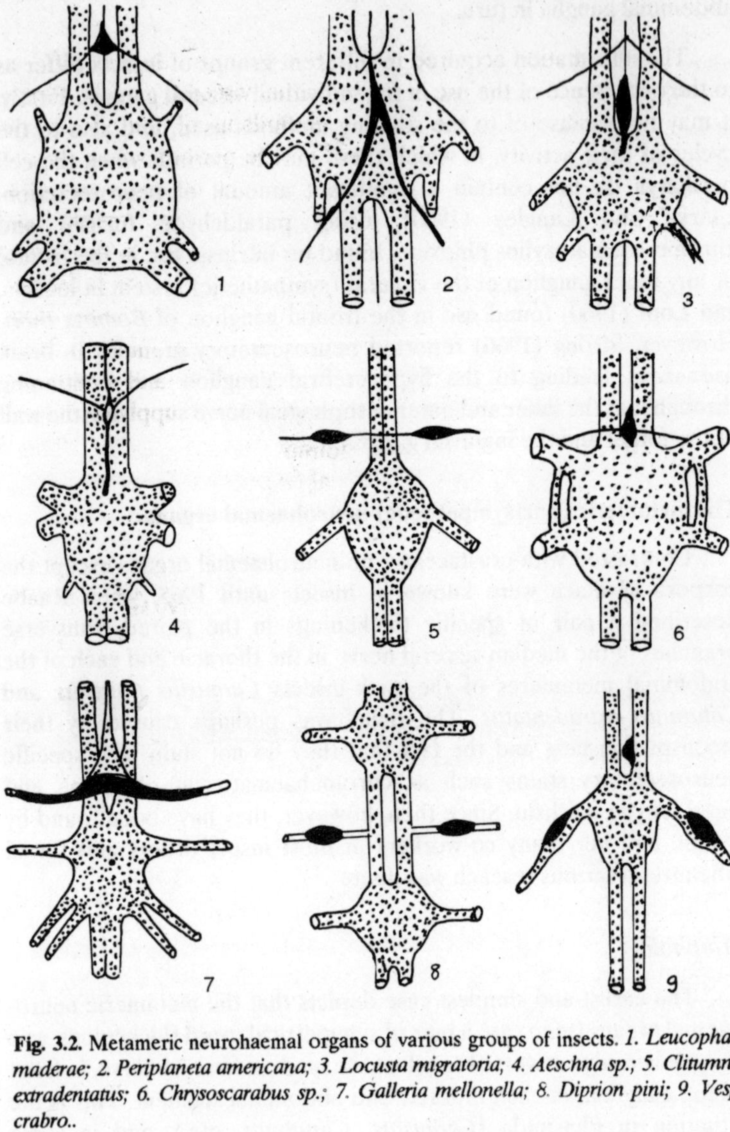

Fig. 3.2. Metameric neurohaemal organs of various groups of insects. *1. Leucophaea maderae; 2. Periplaneta americana; 3. Locusta migratoria; 4. Aeschna sp.; 5. Clitumnus extradentatus; 6. Chrysoscarabus sp.; 7. Galleria mellonella; 8. Diprion pini; 9. Vespa crabro..*

Leucophaea, Periplaneta). Later, the inigtially paired structures may amalgamate to form a single ovoid swelling of the median nerve, near its gangliar attachment and behind its branching point, as in *Locust* or, in a different form, *Aeschna*.

More complex structures of the similar types are found in higher orders, such as the bilobular organ of Lepidoptera, e.g. *Galleria* or the speherical median organ of some Coleoptera. In Hymenoptera, the visceral nerve tends to remain within the central nervous system (the connectives as well as the ganglia). Here the mno arise from the connectives or from the wall of the ganglion (as in *Diprion*), or they may appear as longitudinal thickenings along the anterior surface of the main lateral nerves. In this case the median visceral nerve seems to branch within each ganglion, so that this part is connot be out wasdly seen

In nerves dissected in physiological saline, the neurohaemal organs can be distinguished by their bluish while, opalescent colouration, which is similar to that of the corpora cardiaca. This undoubtedly not because of the actual neurosecretory substance, but because of the pterine accompanying it.

Histology

Histologically, any mno consists of characteristics neurohaemal-tissue. The middle part contains ramified neurosecretory fibers surrounded-by glial cells. The entive structure is enclosed in a thin surface membrane, often with deep invaginations, which is doubtless a secretory product of the glial cells. At times it includes individual nerve cells and neurosecretory cells, as in some of the organs of Hymenoptera, Heteroptera and Coleoptea even though these cells are absent in many other orders.

The transportation of the neurosecretion to the organs along the neurosecretory axons and in most cases the greater part accumulates in the surface layer of the organ in the form of floccules or droplets. It is completely Gomori and paraldehyde-fuchsin negative and stains more readily with azocarmine. It also stains, though less readily, with phloxine. The fact, that it stains deeply with bromophenol blue, Glenner and Lillie's dimethylaminobenzaldehyde-diazo-S acid and fast green FCF indicates that it probably contains proteins with pyrrole or indole. The last two stains mentioned do not act on the phloxinophilic surface floccules, however. As insisted by Raabe and Monjo (1971), it should be noted that the neurosecretory substance in the metameric organs, like that in type C neurosecretory cells, is not visualized after fixation with Bouin's fluid, but is clearly visible after fixation with plain formalin or with Helly's fluid.

It has been asserted by Raabe (1965) that the products of the neurosecretory cells of the ventral ganglia were stored in the mno and passed into the haemolymph. This undoubtedly refers to type C neurosecretory cells, which are rather abundant there however, the accuracy of such assertion us questioned by the findings of Finlayson and Osborne (1968, 1970), who showed in *Carausius* that several neurosecretory perikarya occurred along the link nerve, i.e. a thin nerve branch connecting the trnasverse nerve, together with its neuroshaemel organ, with the main segmental nerve of the particular ganglion. These perikarya corres- pond to the number of axons forming the transverse nerve. The further research deals primarily with the conclusive solution of this problem.

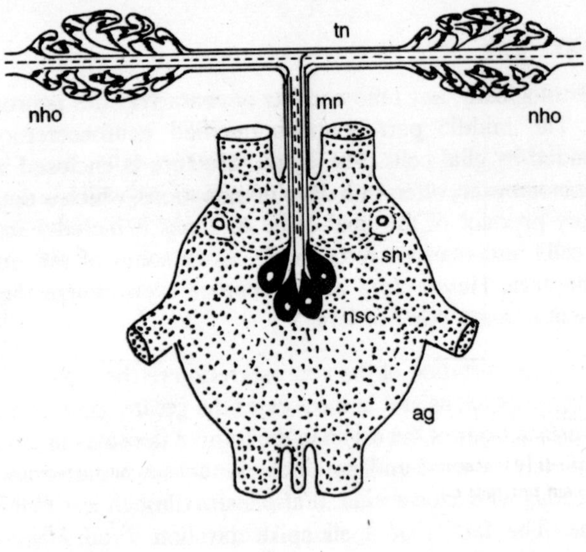

Ultrastructure

An elaborate electron microscopy study of mno in Phasmida (*Carausius* and *Clitumnus*) by Raabe and Ramade (1967) fully confirmed their neurohaemal character. They consist of neurosecretory fibers and their relatively thick surface membrane has large numbers of deep invaginations constituting a genuine internal stroma. The passage of the neurosecretory substance into the haemolymph probably takes place in their many sinuses, which have a very thin surface membrane. Several types of glial cells, vacuoles and lipid and glycogen granules, etc., were found. The ultrastructural appearance of the two types of mno in Hymenoptera (*Vespa* and *Vespula*) is similar. The lateral longitudinal organs attached to the main segmental nerves in this group differs from the median neurohaemal organs, as they have no surface glial membrane and practically no glia cells. They are packed with neurosecretory nerve fibre endings and have large numbers of sinuses. The observertion consists of specific neurosecretory cell containing small granules.

Physiology

Almost nothing, is known of the physiological significance and function of the mno in the insect organism. The primary cause is the impracticability of completely removing such a complicated, ramified structure from the living insect. Some data were obtained by injecting extracts into insects during an inactive period, and by incubating explanted tissues, e.g. heart, Malpighian tubules and rectum, in such extracts. It was demonstrated by these methods that the extracts act upon cardiac rhythm and upon tanning of the cuticle. While the physiological colour changes and the trehalose content were unaffected. Changes in the diurnal activity of the type C neurosecretory cells supposed to deposit their secretion here, indicate that the activity of the durnal rhyth has been interfered with.

Electrical activity

Automatic electrical activity, in the form of action potentials, was demonstrated in *Carausius morosus* mno by Finlayson and Osborne (1970). The action potentials resembled those observed in other neurosecretory cells, for their spike duration was longer than for ordinary neurones. They were produced by the neurohaemal organs irrespective of the presence of absence of any perikarya in the preparation and thus appear to arise in the organ itself. They may have a bearing on the release of secretion. This would coninside with the earlier observation that electrical stimulation results in emptying of the neurosecretory granules from the neurosecretory cells and corpora cardiaca.

Knowledge about the functioning of these organs can be drawn from the following observations by Raabe (1966-1971). If the median nerve is sectioned in front of a ganglion, reduction of the azan-positive material in the neurohaemal organs of the corres- ponding ganglion is observed, together with intensive accumulation of the neurosecretion at the proximal end of the sectioned nerve. Raabe concluded that the substances contained in the neurohaemal organs originate in the neurosecretory cells of the ganglia. However, it remains to be seen. Wether this is also correct of other C-cells. It definitely does not apply to the type A- and B-cells in these species, since their neurosecretion is Gomori- and PAF-positive and so such material was found in the neurohaemal organs. While Musko and Novak (1973) reported a certain amount of Gomori- and PAF-positive granules in the neurohaemal organs in *Gryllotalpa*.

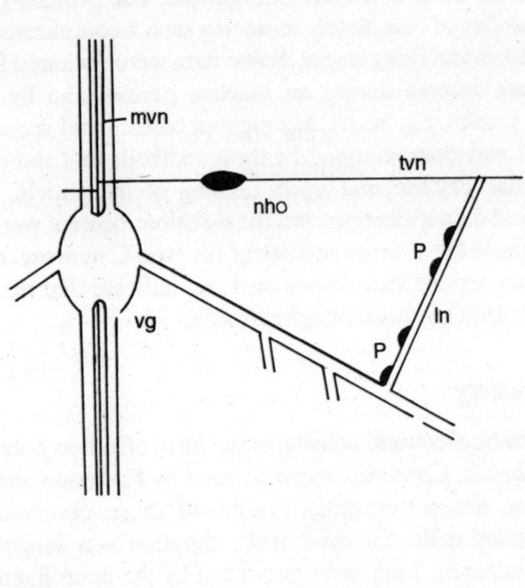

A study was made by Smalley (1970 about the history and ultrastructure of the nsc of the abdominal ganglia of *Periplaneta americana* whose axons enter the paired metameric neurohaemal organ, which has united here to form a single structure in the median line. The concerned cells are divided into the three groups of neurosecretory cells in the ganglion posterior to the organ were found to be concerned: the mid-line median nerve cells situated between the connectives entering the ganglion, the nsc group behind the ventral neuropile and a third group below the centre of the dorsal surface of the ganglion. The cells were negative in specific staining reactions, but took up ^{3}H from injected tritiated dopamine more readily than the other cells of the ganglion. In the neurohaimal organs Electron transparent vesicles were observed in their endings in the neurohaemal organs.

An observation made by Novak (1959) shows a secondary neurohaemal organ, in the form of a thickening of the paired nervus allatus II was studied by Belyaeva (1966) and Theodorescu and Novak (1973) in the same species and by Awasthi (1968, 1971) in *Gryllodes sigillatus*. Each of the paired dilations is spindle-shaped and contains, around the central swollen nerve fibres, large numbers of glia cells covered by a sheath of the same nature as the neural lamella and, in fact, is continuous with it by means of the sheath of the nervus allatus II. The substance which fills the organs is mainly a Gomori-positive neurosecretion with different sized granules, coming from the brain via the nervi corporum cardiacarum, the corpus cardiacum, the nervus allatus I, the corpus allatum and the nervus allatus II. No neurosecretion was observed in the last (and thinnest) segment of the nerve (II D) between the thickening and the suboesophageal ganglion. The thickenings of the nervus allatus II have a regular structure, which the metameric neurohaemal organs.

Musko and Novak (1973) studied the metameric neurohaemal organs in *Gryllotalpa gryllotalpa*, they detected a pair of large neurosecretory cells staining red with azan, but Gomori- and PAF-negative, at the posterior end of each ganglion of the ventral nerve cord, in the median line between the two connectives. The axon of each of these cells eutered the transverse branch of the median nerve on its body side. They formed part of a small group of neurones (10 to 20) surrounded by small cells of a glial character and enclosed the impression which was given by the in a thin sheath staining blue with azan (like the neural lamella). The whole structure of a small incorporated ganglion connected to the median visceral nerve. On the basis of their findings, the authors formulated the hypothesis that this is relic of the original state of the median visceral system, which they suppose originally had a small ganglion in each body segment. Their foundings lead to their belief that the frontal and hypocerebral ganglia are simply specialized forms of similar ganglia of the cerebral and suboesophageal ganglia.

The ganglionic neurohaemal organs

Another important discovery was that of a special pair of neurohaemal organs in each of the ganglia of the ventral nerve cord. They are placed each on a thin paired nerve rising from the laterodorsal region of the ganglion tightly to its surface. These organs are thicker, more spacious and with less deep constrictions than the perisympathetic neurohaemal organs. They seem to drain neurosecretion from nsc of the lateral parts of the ganglia. Unlike the perisymapathetic organs, they contain granules of paraldehyde-fuchsine (Gomori-positive) neurosecretion bedsie of the asian-positive (phloxinophylic) one. The organs have been titled as gangdionic neurohaimal organs.

SURVEY OF NEUROHORMONES

Myotropic effects

Loeuris classic experiment triggred the experiments on the myotropic effects of neuro-hormones are Loewi's classic experiments. These effect may be demonstrated on insect's organs with a spontaneous rhythmic activity such as the heart, Malpighian tubules, oviducts and gut. One of the first authors to show in insects the hormone dependence of the pulsations in isolated Malpighian tubules, oviducts and gut was Koller (1948). He showed that extracts of brains and corpora cardiaca accelerate the contractions of isolated Malpighian tubules, oviducts and gut was Koller (1948). He showed that extracts of brain and corpora cardiaca accelerate the contractions of isolated Malpighian tubules in insects-Ringger solution, wereas the same extracts reduce the peristaltic movements of the gut when the conditions are alike. In contrast, stimulation of gut pusations are caused by extracts of the gut and corpora allata. Enders (1956) has found the opposite effect of extracts from brains, corpora cardiaca and corpora allata on the peristalsis of the oviducts in the stick-insect *Carausius morosus*. The effect of extracts of corpora cardiaca and corpora allata was negative while that of the brain extacts was positive.

Experiments with Corethra. Exhaustive research into myotropic effects in the transparent larvae of *Corethra* (*Chaoborus crystallinus*) has been made by Gersch (1960a, b) and his co-workers. There is a great advantage in this species because the effect of various extracts on the gut can be made visible by the use of various vital stains and can be easily observed in *vivo* owing to the transparency of the larva's body. A characteristic feature of the species is the strong antiperistaltic move- ments which follow each food acceptance. These movements bring about the digestive enzymes into the pharynx, where the process of digestion in this insectivorous species commences. Mechanical of thermal irritation of the brain and of single ganglia in the ventral nerve cord has demonstrated the dependence of these antiperistaltic movements on specific ganglia. The three indi- cations in favour of a humoral character for this effect: are Firstly (1) The effect of the irritation only appeared after a definite 'late' period; it subsequently increased and continued to affect the movement for about one hour. (2) Secondly he observed effect was in no way reduced when the adjacent ganglion on wither side had been destroyed by burning so that the connectives were interrupted. (3) Thirdly no nervous branches entering the gut walls could be found arising from these ganglia, the only nerve supply to the intestine being the stomatogastric nervous system arising from the brain. A similar effect was obtained by irritating the cerebral ganglion, with the only difference that in this case, the rise in the rate of peristaltic movement was preceded by a wave of increased activity of short duration. Undoubtedly such an effect was because of a nervous impulse from the stomatogastric system. The proof of a humoral action of the ganglia via the haemolymph was obtained by experiments with extracts, not only from the various ganglia of the nervous system of the same species (*Chaoborus*), but also from ganglia of the cockroach *Blatta orientalis* and the stick-insect *Carausius morosus*. These extracts were equally active on the isolated gut in insect-Ringer. The relevant active substances were obtained from all parts of the nervous system, and were specific neither for species nor order. Consequent research has shown that they affect the heart-beat not only in other arthropods but even in other animal Phyla such as molluscs and vertebrates.

A humoral factor obtained by homogenization of the corpora cardiaca

Another research shows that this substance stimulates the pericardia cells to produce a heart-accelerating principle in *Periplaneta americana*, this was described by Davey (1961, 1962) in greater detail. Pericardial cells treated with the factor *in vitro* show histological signs of increased secretory activity and produce a substance increasing the heart rate, which was identified as indolalkylamine. Where the live animals are concerned, the release of this factor from the corpora can be induced by feeding. As suggested by Gersch (1964) there are no grounds for assuming that this form of interaction of the pericardial cells is general in insects, and there is good evidence of the direct effect on the heart rate of the neurohormones isolated, amongst others, from the corpora cardiaca. As emphasized by Evans (1962) and Gersch (1964), it would be very interesting to determine the relationship of Davey's corpus cardiacum hormone' to the neurohormones C_1 and D_1 and possibly to the other active principles earlier demonstrated to have happened in the corpora cardiaca as well as to the secretion of the pericardial cells.

The isolation of neurohormones

The effects which are mention above were investigated by Gersch's co- worker Unger (1956), who used isolated hearts of *Blatta orientals* and *Periplaneta americana* as test material. Insect-Ringer extracts from various parts of the nervous system were used. It is known that the heart of insects is innervated partly by the stomatogastric system and partly by thin nerve branches from the corresponding abdominal ganglia. The pulsation may therefore be influenced directly by either of the nervous paths. It was shown in the insect- Ringer preparations that the heart-beat could be stimulated by means of the same extracts as had been found to influence gut peristalsis. In contrast, extracts of the corpora allata subdued the heart-beat. Lot of uncertainty has been seeen in this effect the cockroach heart preparations were influenced not only by extracts from cockroaches but also by extracts from stick-insects, dragonfly larvae and blowflies (*Calliphora*).

Comprehensive examination of such extracts and their separation by paper chromatography has disclosed that no less than three physiologically different substances can be isolated. One of these substances, one is discovered in all parts of the ventral nerve cord but missing in the brain and the retrocerebral glands, was shown to be identical with acetylcholine. It stimulated the heart-beat and gut peristalsis. The other substance occurring in the brain and corpora cardiaca as well as in the ventral nerve cord was called neurohormone D by Gersch and Unger (1951). This equally stimulated the frequency of the heart-beat but at the same time increased its amplitude. At high concentrations it stopped the heart-beat during diastole. The distinctive fluroescence was later explained to be because of a mixture of pterines which are very hard to separate. A third substance, partially antagonistic to neurohormone D and denoted neurohormone C by the same authors, was found in all parts of the nervous system and in the corpora cardiaca like neuro- hormone D, but was also found in the corpora allata. It too increased the frequency of the heart-beat, but decreased its amplitude, and at higher concentrations, stopped the heart during systole. In this respect the effect of neurohormone C is antagonistic to the effect of neurohormone D. The haemolymph was isolated from thses three substance.

Advanced research on the same theory enabled Gersch and his collaborators to separate the original neurohormone D into two different substances D_1 and D_2 and the original C into hormones C_1 and C_2. These were extracted from 3200 lyophilized central nervous systems of *Periplaneta americana* with a succession of petroleum other, ethyl ether, ethyl acetate, ethanol, water, acetic acid and pyridine. The largest amount of the physiologically active component was obtained from the water fraction, and was found to be identical with the ethyl alcohol fraction. Paper chromatography of the water-soluble fraction with butanol-pyridine-water (1 : 2.5 : 2.5) clearly showed two substances: one with the activity of neurohormone C (at an RF of 0.17) and the other with the activity of neurohormone (at an RF of 0.33). They were eluted with water and shown by bioassay to correspond with the previously described hormones. The substance with the physiological action of neurohormone D was recrystallized several times to yield about 50 μg. of fine spheroid crystals with a fairly constant form. This substance was designated as neurohormone D_1 and some of its physical properties were determined. The absorption peaks in the infrared were cfound at 1140, 1400, 1480, 1640, 3400 and 3600 cm^{-1}. In this relatively pure form it was slightly soluble in water, hardly soluble in ethyl alcohol, and insoluble in organic solvents. When peptides are added, however, it readily dis- solves in water. The second substance recrystallized in needle form, showing a tendency to fuse, and corresponded with neurohormone C in physiological activity and was designated as neurohormone C_1.

Another two substance which resemble and yet are distinctly different features were obtained from the ethyl acetate fraction by paper chromatography with butanol-ethanol-acetic acid-water. One of them, with an RF of 0.17, was designated as neurohormone C_2 and agreed with C_1 in its positive effects on the frequency of the heart-beat without inducing systolic contraction. The other substance D_2 with an RF of 0.25, corresponded (approximately) with D_1 in its crystal form, and showed similar stimulation of the heart-beat. Unlike C_1 and D_1, neither C_2 nor D_2 produced any effect on colour change. The possibility must be taken into account that they may be either decomposition products or intermediate compounds in the synthesis of C_1 and D_1. The four substances could not be identified with either adrenaline, noradrenaline or histamine (Unger, 1956). While a series of substances of synthetic or vertebrate origin have proved to be active in isolated gut or heart experiments. For example acetylcholine, doryl, physostigmine and histamine have been found to stimulate antiperistaltic movements, the most active being acetylcholine which is still active at a concentration of 10^{-7}. In contrast, the effect of adrenaline and pilocarpine is distinctly negative, damping the movement. Extracts of the nerve ring and the ventral ganglion of Ascaris affect the heart-beat of the snail *Helix pomatia*. Similarly, extracts from the snail *Alpysia* affect the heart-beat of *Chaoborus*. This demonstates that the neurohormones are remarkably non-specific.

Exhaustive research made by Gersch (1962) was on the possible identity of some of these neurohormones with the AH produced by the medial neurosecretory cells of the brain, using last instar larvae of *Calliphora erythrocephala* and *Lymantria despair*. A series of experiments involving the injection of brain extracts, injection of the isolated neurohormones C_1, D_1, C_2, and D_2, as well as implantation of immature ring glands suggested that neurohormone D_1 was identical with AH. It was possible to induce pupation in brainless *Lymantria dispar* larvae by injecting the purified neurohormone D_1. Similar results were obtained in *Calliphora*. The neurohormone C_1, however, showed no such activity either in *Lymantria* or,with two or three exceptions, in *Calliphora*.

Comtemporary research using electrophoretic sepa- ration of brain extracts on polyacrylamide gels (Sephadex G- 100) has proved that AH is different from neurohormone D, despite the fact that they are both of a peptide nature. Neurohormone D was found to be an oligopeptide with a molecular weight of about 2000, with specific biochemical properties (thermolability, acid lability, quick decomposition by trypsin, etc.). SH groups are an important component of its molecule, as they are essential for its action. Therefore it is somewhat to the vasopressin of vertebrates. In addition to the above properties, it was shown to reduce significantly the membrane potential of the prothoracic glands, while activation hormone I had no effect, and activation hormone II raised the membrane potential. It was active in vertebrates as well as in spiders and crustaceans, affecting water flux and Na transport across the wall of the urinary bladder of the toad in the same way as vasopressin.

There are six factors which have an accelerator action on the isolated heart of *Periplaneta* were extracted in ethyl acetate, ethyl acetate-ethanol mixtures, and ethanol from the corpora cardiaca-allata of the cockroach by Ralph (1962). Another accelerator, soluble in saline solutions, was found in all neural structures examined. It is possible that five factors caused heart deceleration were found by the same solvent extrication authored in the brain, and the suboesophageal and ventral cord ganglia. Almost nothing is known about the actual function of the myotropic neurohormones in he insect organism. It can be assumed, however, that an increase or decrease in heart rhythm and peristaltic movements is of adaptive value for insects under certain specific conditions. The mechanisms of any such correlations still remain to be determined. Another function could be the induction of specific processes, such as ecdysis, by the ingestion of air or water, fledging movements, etc., or of oviposition, with the necessary contraction of the oviduct and abdominal body wall. The latter can be deduced from papers like those by Enders (1955), Davey (1965, 1967), de Besse (1965). Thomas and Mesnier (1973) analysed this mechanism in stick- insects (*Carausius morosus* and *Clitumnus extradentatus*). Two types of neurosecretory factors were shown to be concerned with egg-laying in these species. One is a stimulant sub- stance produced by A-nsc in the cerebral ganglion and the suboesophageal, thoracic and abdominal ganglia, while the other (of an inhibitory character) has been localized solely in the brain. The experiment shows that these two factors seem to combine to induce the circadian rhythm of egg-laying in these insects.

Rhythms of activity

Number of papers have proved that there is a causal connection between cyclical, seasonal and diurnal activity and the activity of the neurosecretory cells. Klug (1958) has shown that there is a correlation between winter and summer diapause and the volume of the nuclei of the corpora allata. It has also been seen that the neurosecretory cells of the pars inter- cerebralis were found to be full of secretion at noon, but were almost empty at 4 o'clock in the morning.

It has been pionted by Oezbas and Hodgson (1958) have shown that an extract of the corpora cardiaca decreases the 'spontaneous' electrical activity in isolated ganglia of the ventral nerve cord in *Periplaneta americana*, whereas extracts of the brain, abdominal ganglia and corpora allata have no effect. Cutting the nerve between the brain and corpora cardiaca made extracts of the latter less effective. When the corpora cardiaca extracts were injected into living animals, movements became un-coordinated and stereotyped locomotor behaviour appeared several hours after the treatment and remained for about the span of sone to three days.

A research was by Hatker on the endocrine control of the activity rhythm of the same species, using parabiosis and transplantation experiments. Her results showed that the suboeso- phageal ganglion was the source of the active substance involved. Injecting extracts of this ganglion into decapitated insects produced a period of increased activity, and the time of day when the injection was carreed aou is decisive factor in the consideration of the duration of the effect.

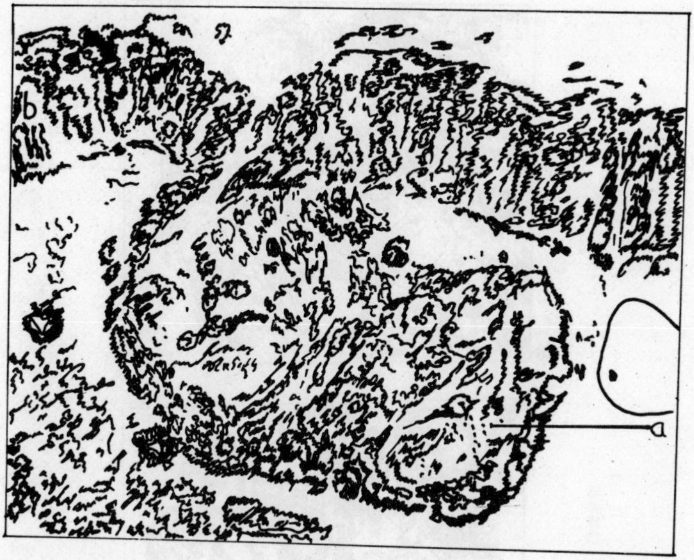

(a)

Fig. 3.5. (a) Hypertrophy of the gut wall of *Carausius morosus* following allatectomy in an early larvel instar, a - cancer-like growth of the gut wall; b - gut epithelium. (b) Hypertrophy of lymphocytes following allatectomy in *Carausius morosus*.

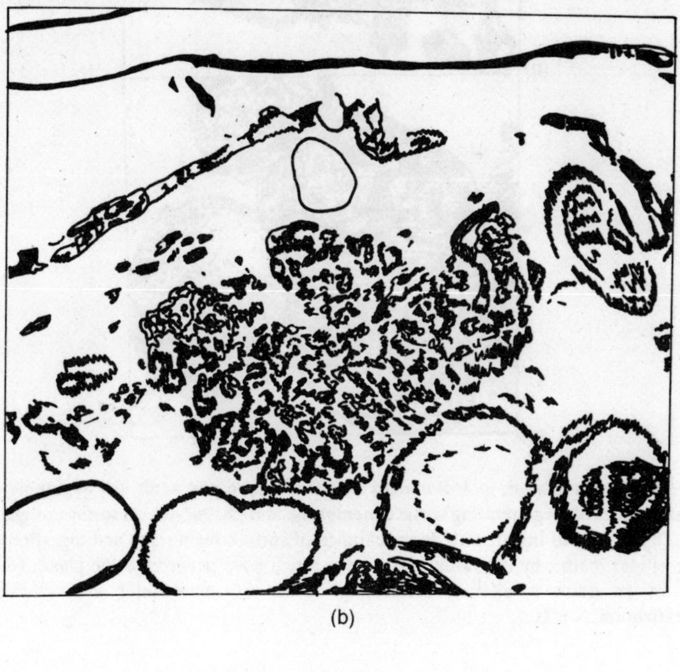

(b)

(a)

(b)

(c)

Fig. 3.6. Corpus allatum in *Mammestra brassicae* pupae and adult. (a) beginning of pupal instar, with degenerating surface membrane and phagocytes on surface of gland only, (b) 4th day of instar, with disappearance of surface membrane and digestion of intercellular matter by individual phagocytes which have penetrated the gland, (c) 4 cells of an adult c, allatum completely devoid of intercellular matter, same magnification. × 800.

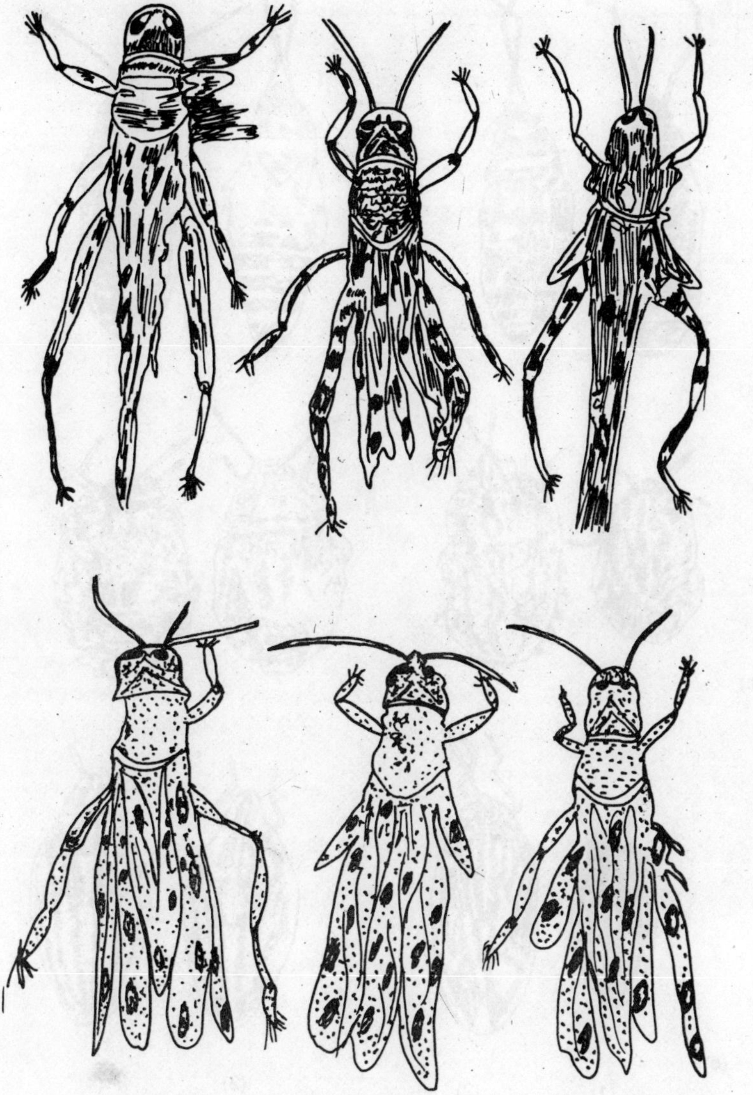

(d) (e) (f)

Fig. 3.7. Intermediates between normal adult (a) and Vth (last) instar nymph (c) of *Schistocerca gregaria*. (b, d, e, f) intermediates of varying degrees, produced by topical application of juvenoid. Note the larval shape and coloration of the pronotum in the intermediates.

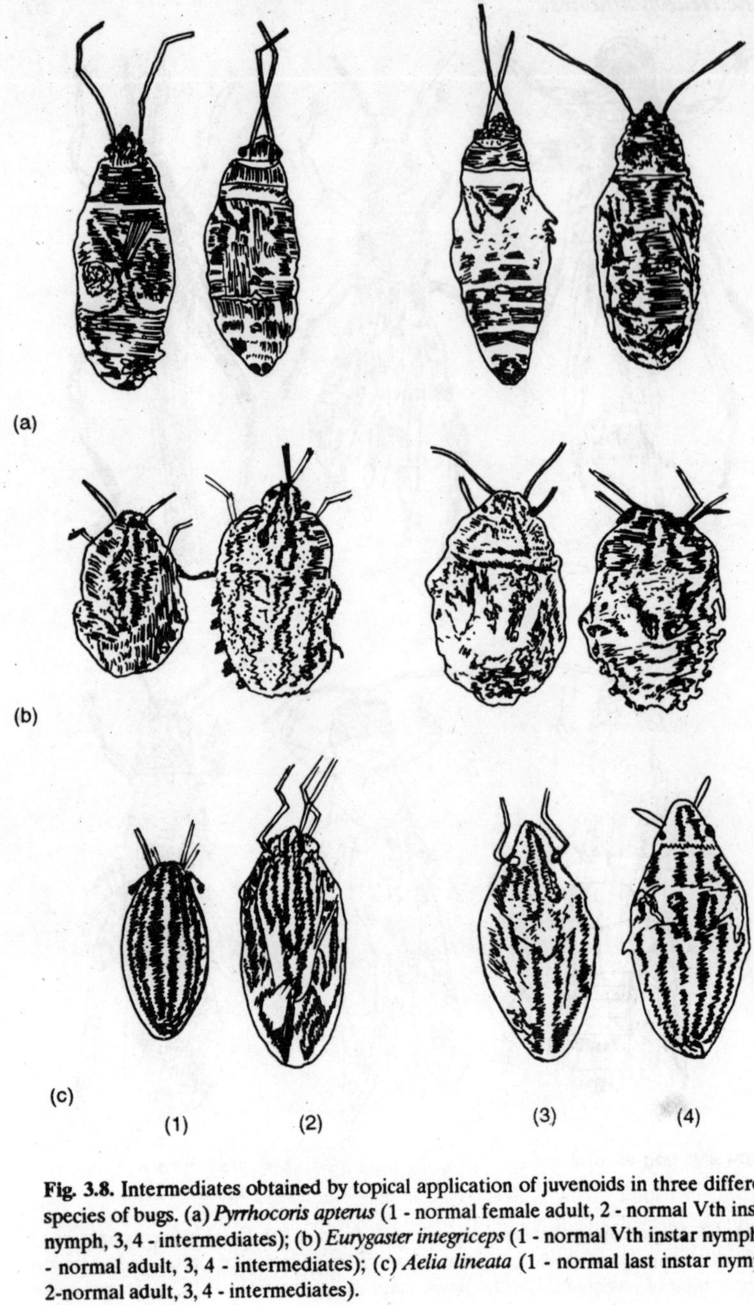

Fig. 3.8. Intermediates obtained by topical application of juvenoids in three different species of bugs. (a) *Pyrrhocoris apterus* (1 - normal female adult, 2 - normal Vth instar nymph, 3, 4 - intermediates); (b) *Eurygaster integriceps* (1 - normal Vth instar nymph, 2 - normal adult, 3, 4 - intermediates); (c) *Aelia lineata* (1 - normal last instar nymph, 2-normal adult, 3, 4 - intermediates).

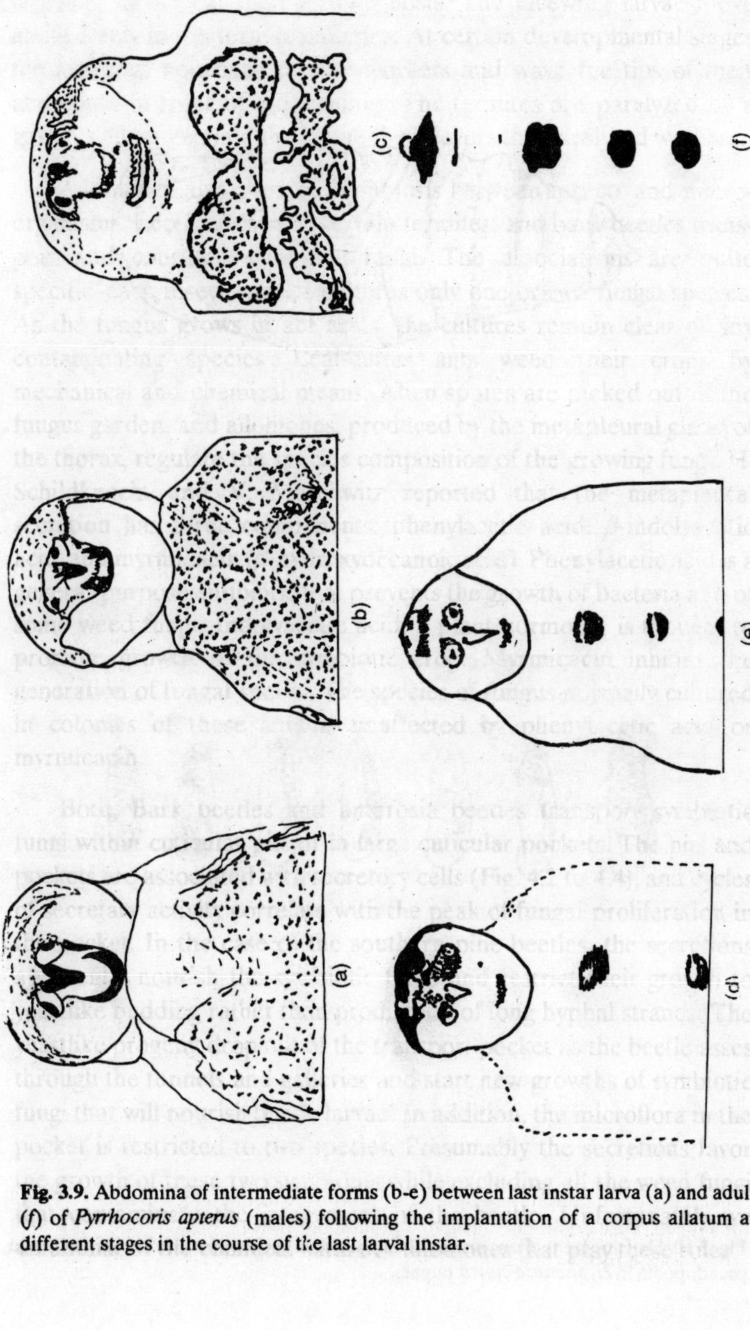

Fig. 3.9. Abdomina of intermediate forms (b-e) between last instar larva (a) and adult (*f*) of *Pyrrhocoris apterus* (males) following the implantation of a corpus allatum at different stages in the course of the last larval instar.

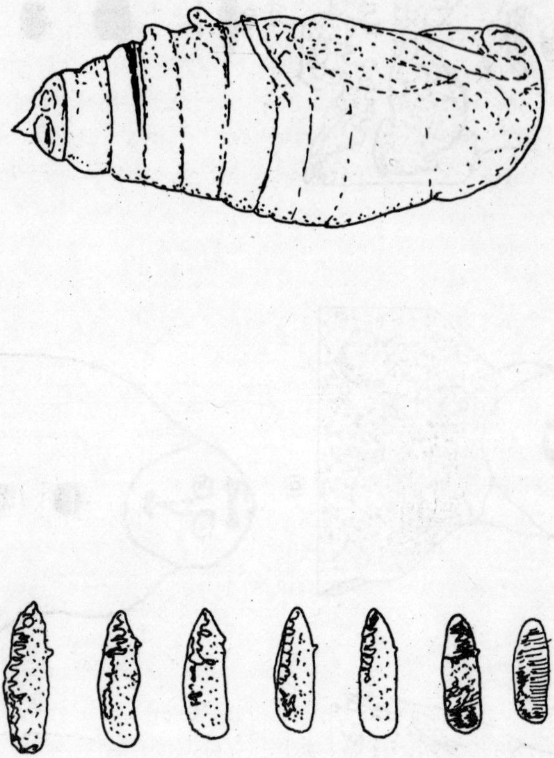

Fig. 3.10. Development of anomalies of a hysterothetelic nature. The remains of larval pseudopodia in *Hyphantria cunea* pupae.

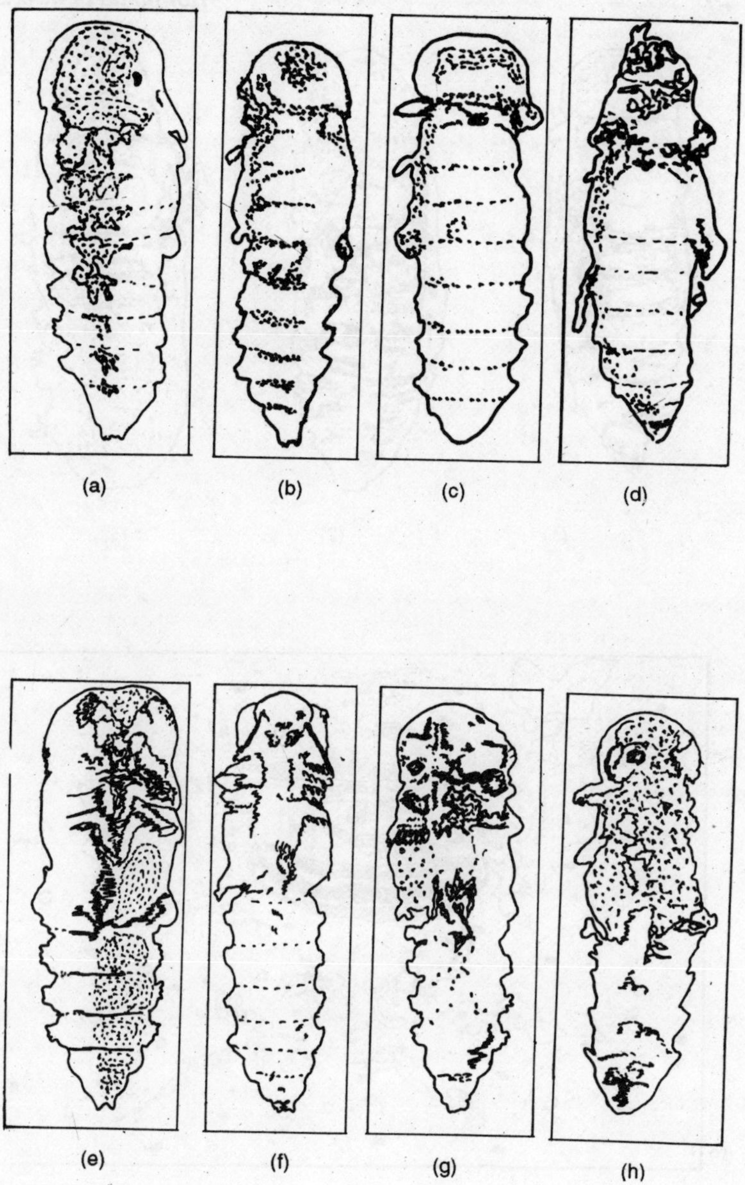

Fig. 3.11. *Tenebrio molitor* pupae affected by topical juvenoid treatment. (a-d) dorsal views, (e-h) ventral views.

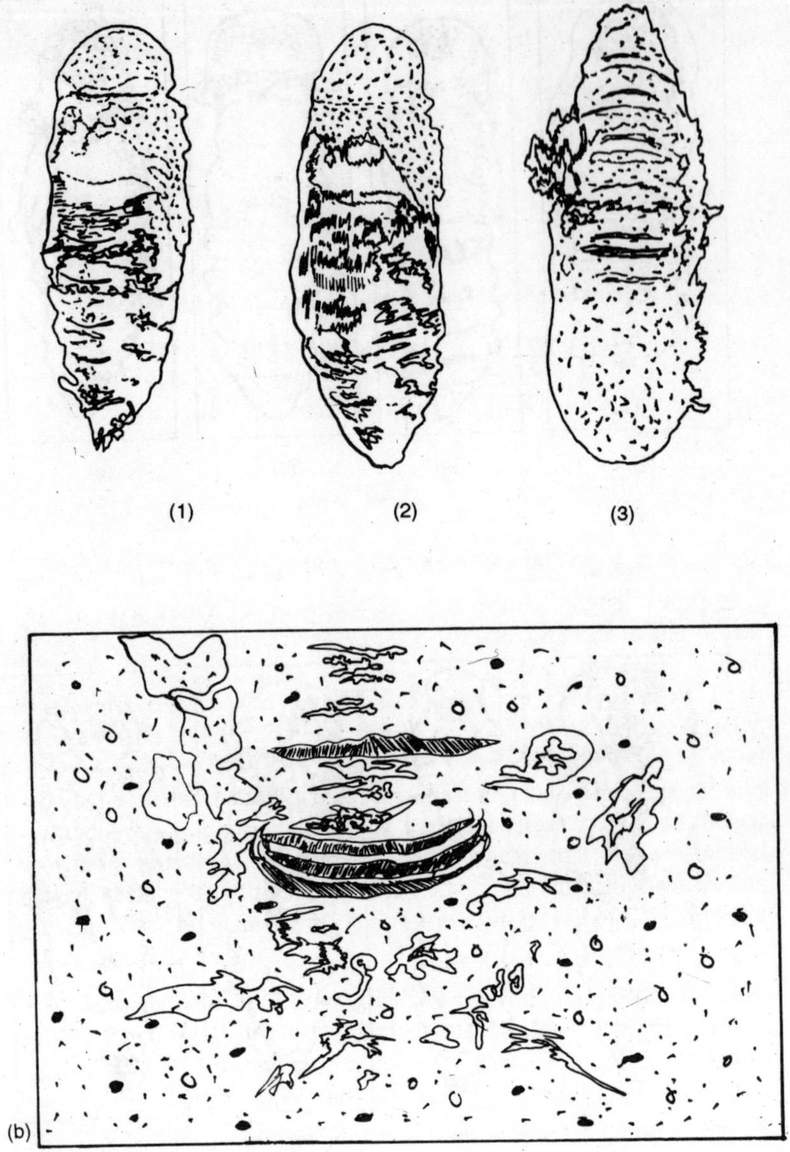

Fig. 3.12. *Dermestes vulpinus* pupae affected by Juvenoid treatment. (a) 1 - normal pupa, 2 - second pupa, 3 - third pupa. (b) three last abdominal terga of third pupa. Note the three chitinous ridges on the last tergites, which are specific for the double pharate pupa.

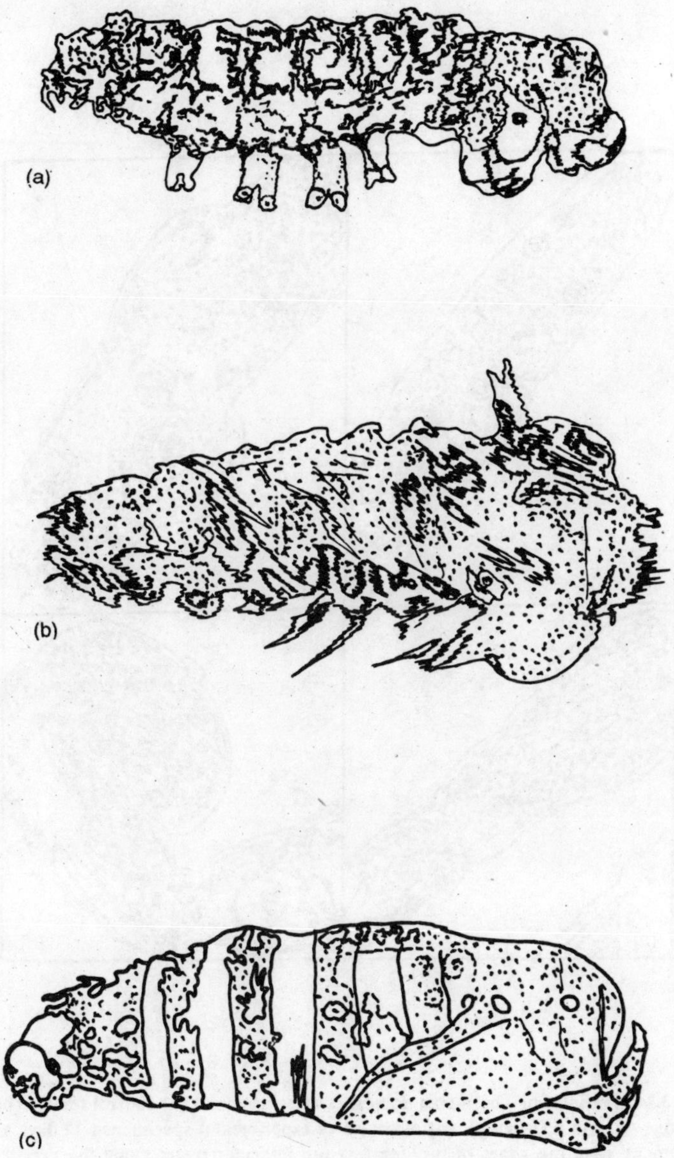

Fig. 3.13. Three lymantria dispar larvo-pupal intermediates produced by juveniod treatment. (a) most strongly aggected, (c) least strongly affected.

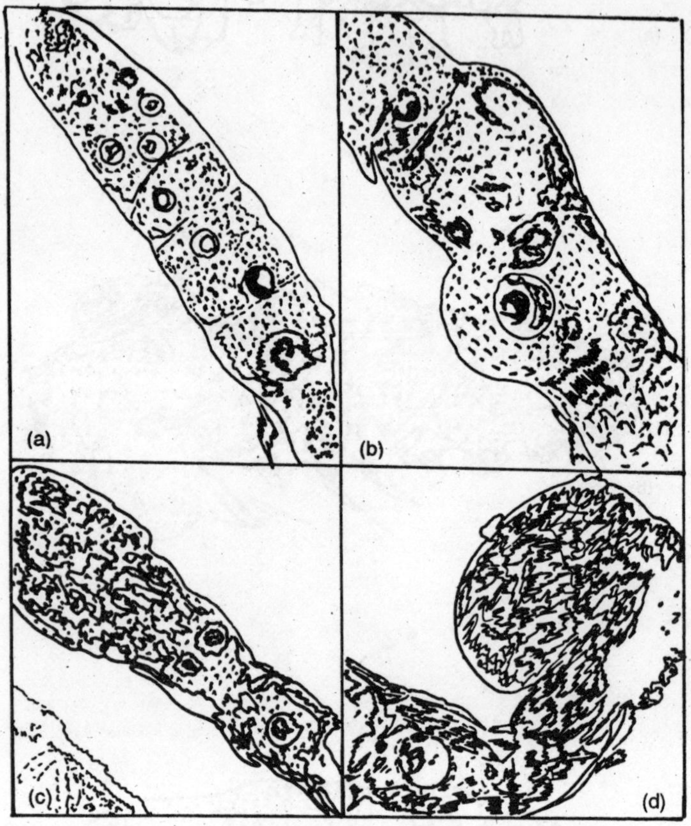

Fig. 3.14 Ovarioles of Thermobia domestics after juvenoid. (a) control female (plain acetone 42 days proviously), (b) ovarioles of experimental specimensn 42 days after treatment; note the zones of proliferating pre follicular tissue along the previtellar oocytes, (c,d) the same, showing varying degreesof hypertophy of the germarium. Living tissue, phase contrast. x 320.

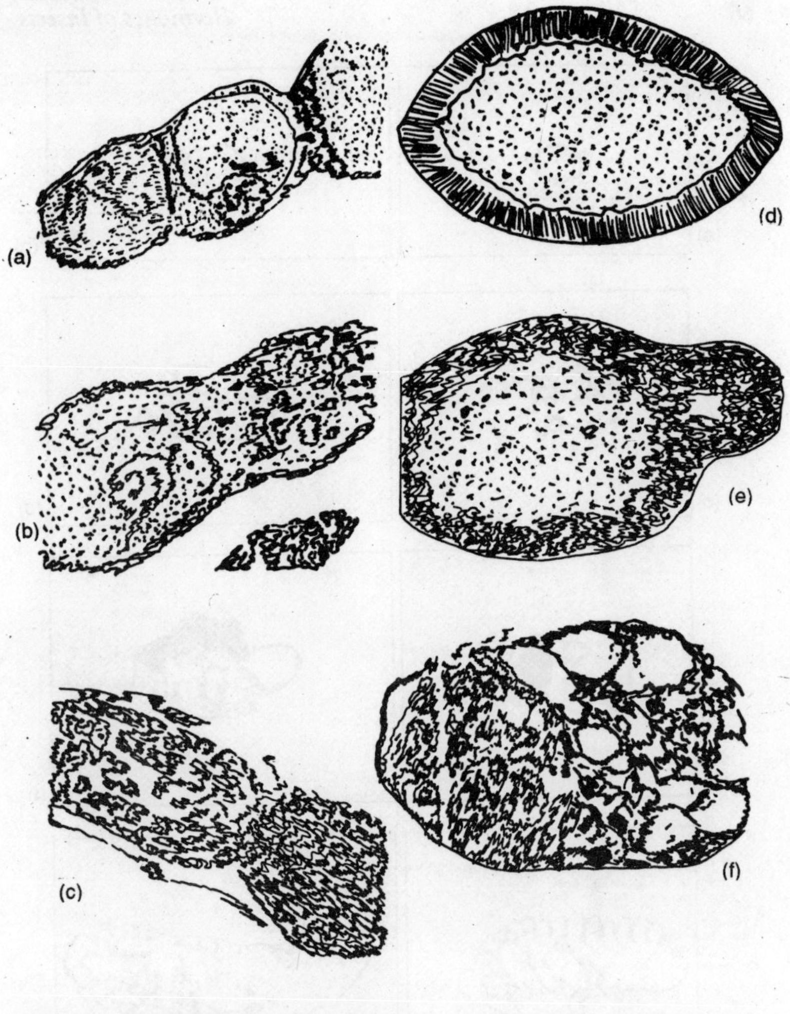

Fig. 3.15. Sections of ovarioles of *Dixippus morosus* specimens after juvenoid treatment. (a) part of vitellarium of intermediate produced by applying JHa to last instar nymph (6 days after moulting, 8 days after application); one follicle with two nuclei, involvement of follicular zone. (b) more severely affected specimen, with obliteration of interfollicular epithelium. (c) section of germarium and part of previtellarium of 20-day last instar nymphs treated on 15th day after ecdysis; both normal and injured oogonia. (d) distal follicle in late previtellogenic stage in 8-day nymph treated immediately after ecdysis. (e) severely affected distal follicle of nymph 14 days after ecdysis (treated between 2nd and 12th day after ecdysis). (f) terminal phase of resorption of distal follicle in 1-day-old abnormal adult obtained from supernumerary nymph.

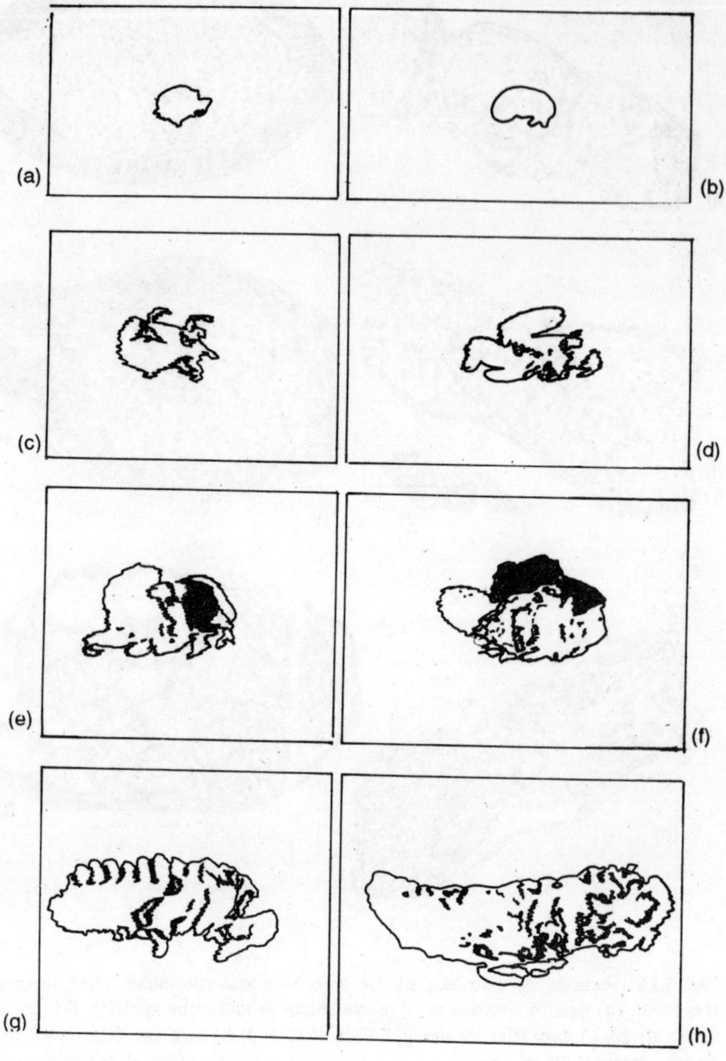

Fig. 3.16. Juvenilized *Schistocerca gregaria* embryos dissected from eggs after treatment with farnesyl methyl ether and fixed at time when controls were hatched. (a) apodous stage with discernible eye pigment, (b) protopod stage, (c) about half way through blastokinesis, (d) second half of blastokinesis (note marked sclerotization and pigmentation as distinct from normal specimens of same age, see Fig.), (e) miniature embryo at end of blastokinesis; (f) the same stage, slightly later, embryo larger and of almost normal size, with pigmented pleuropodium, but with stunted appendages.

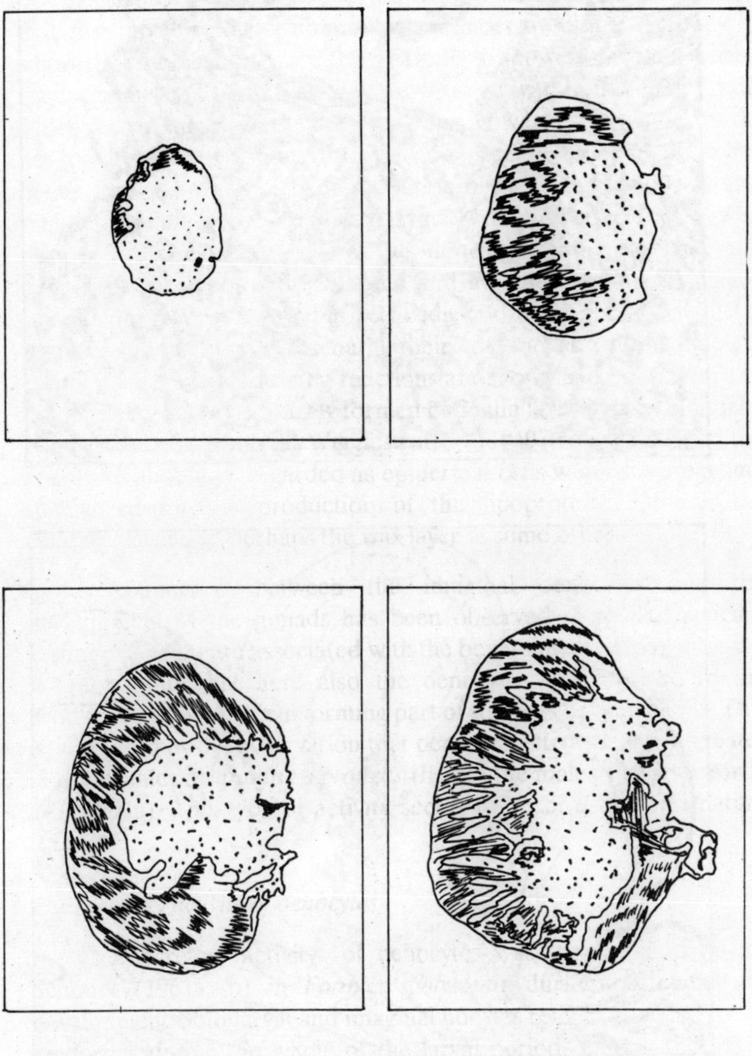

Fig. 3.17. Imaginal discs at four successive stages of development of the last larval instar of *Galleria mellonella*.

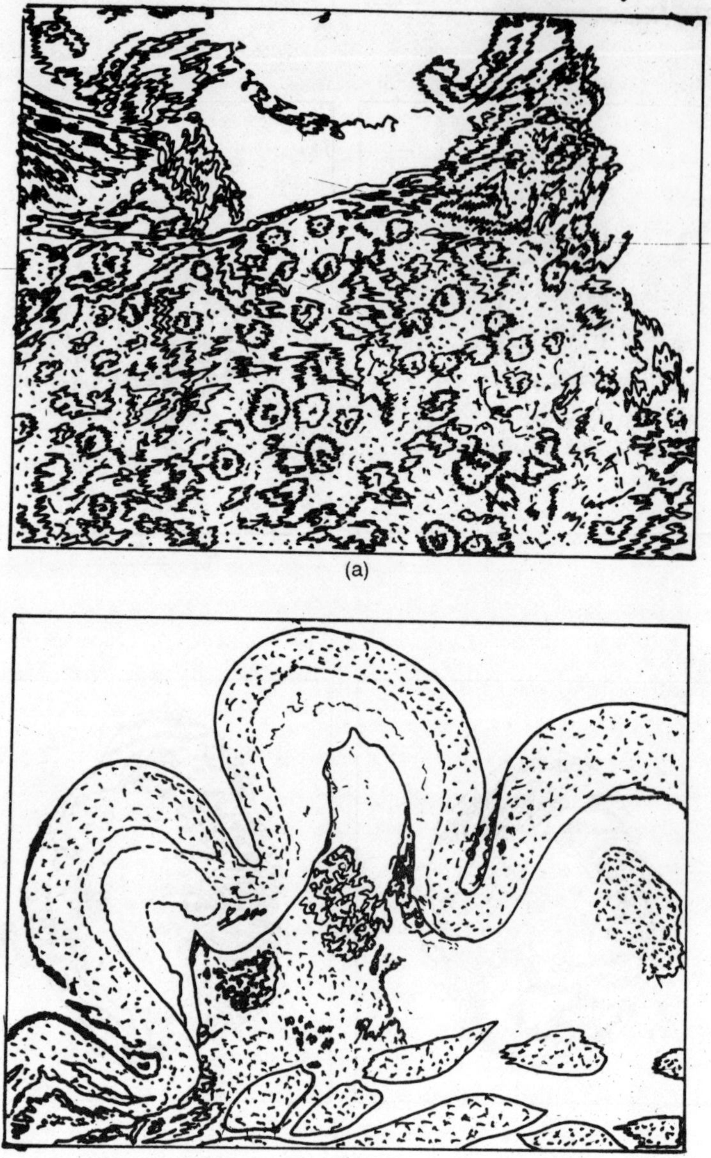

Fig. 3.18. (a) Part of a median section through the corpus allatum of an adult female *Oncopeltus fasciatus* stained with Dobell's ammonium molybdenate-haematoxylin method.. (b) Sagittal section through the epidermis of a diapausing pronymph of *Cephalcia abietis*.

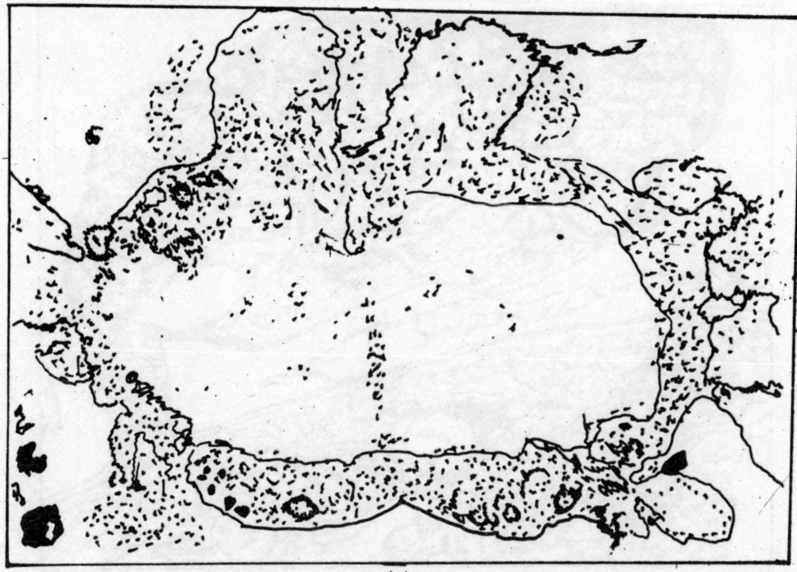

(a)

Fig. 3.19. Gliosecretion in *Periplaneta americana*. Xth instar nymph. (a) Frontal section of the mesothoracic ganglion. Arrows indicate the main accumulations of the gliosecretion. 120: 1. (b) Detail from the above, 1500: 1. Interneuronal gliosecretion around gliocyte nuclei and trabeculocytes with trabeculae.

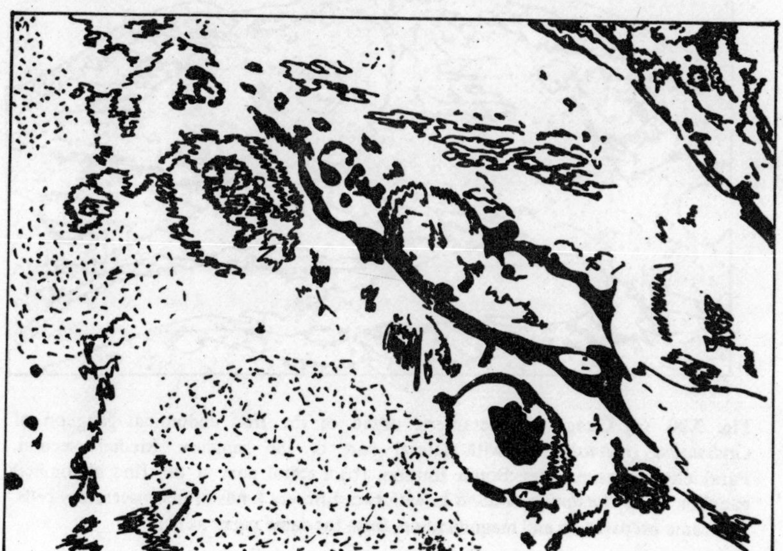

(b)

(a)

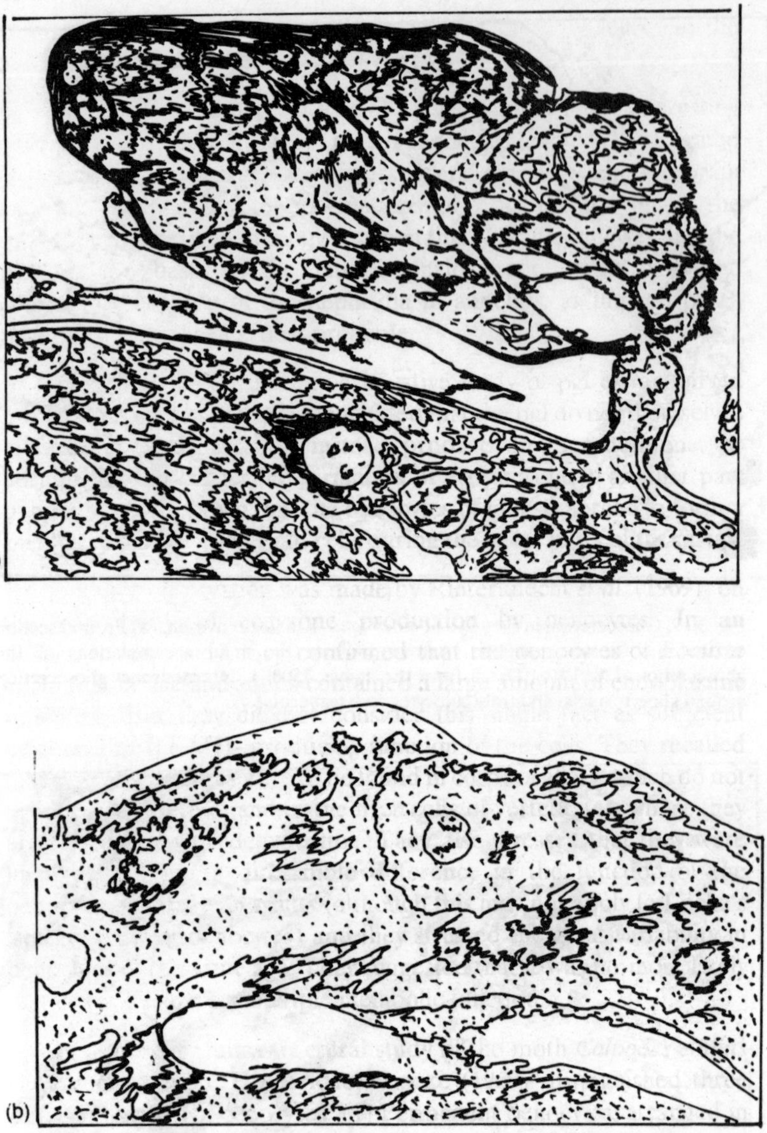

(b)

Fig. 3.20. (a) Ganglionic metameric organ of the IInd abdominal ganglion of *Gryllotalpa gryllotalpa* L. with lateral part of the ganglion. Median section. Paraldehydefuchsine after Bouin fixation. (b) Lateral part of the IInd abdominal ganglion of *Gryllotalpa gryllotalpa* L. with two different kinds of neurosecretory cells. The same preparation and magnification, from the same paper as (a).

The role of neurosecretion and the interaction of other parts of the endocrine and nervous system in controlling circadian rhythms of activity

A detailed study was carried out Harker (1960a, b, 1961) in *Periplaneta americana*. Immediate regulation is effected by cyclical release in a circadian rhythm of secretion by paired neurosecretory cells situated on the ventral and slightly lateral surface of the suboesophageal ganglion. The neurosecretion which induces this phase of activity is produced daily in correlation with the onset of darkness. However, this dependence is not autonomous. Perhaps some other substance maintains their function; this second substance comes from the corpora cardiaca and enters the suboesophageal ganglion via the corpus allatum-suboesophageal ganglion nerve, in the form of methylene-blue-positive granules. Once induced, the rhythm of neurosecretory activity continues for a number of days,inspite of the broken nervous connections of the ganglion.

Midgut tumours were produced artificially by the same author (Harker, 1958, 1960b) in *Periplaneta americana*, by experimentally breaking the above rhythms in an experiment. Transplantable metastasing tumours were induced by the repeated implantation of suboesophageal ganglia out of phase with the animal's own ganglion and by breaking the neurosecretory cycle through specific environmental light conditions. It has been suggested that the production of tumours by recurrent nerve severance, as shown by B. Scharrer (1945), is also related to the timing of suboesophageal ganglion secretory activity (Harker, 1960b).

Another entomologist who carried out an exhaustive study is Mothes (1960) on diurnal activity and the rhythm of colour change in *Carausius morosus*, using the *in vitro* colour change test already mentioned. The quantities of neurohor- mones C_1 and D_1 in Ringer extracts of the brain, the suboesophageal ganglion, the ventral nerve cord, the corpora cardiaca and corpora allata, and the haemolymph were calculated. The resuts of there experiments show that the concentration of both hormones in the haemolymph was very high at night and low by day Gersch (1960b) considers-his earlier view, that neuro- hormone C caused darkening and neurohormone D lightening of the body, to have been disproved. Rather the period of decreased activity and light cuticle is when the quantity of the neurohormones in the haemolymph is low, and there is the period of high activity and dark cuticle during the night.

Brain extracts prepared from the subject at the end of the night (early in the morning) showed low activity, whereas the activity of extracts increased during the day.

Extracts of the suboesophageal ganglion and the ventral nerve cord demonstration the same activity (high at night and low by day) as the haemolymph, whereas extracts of the corpora cardiaca and corpora allata showed no diurnal change in activity.

Reversed conditions of day and night for many days caused the animals to change their rhythm of activity, the activity of extracts also became reversed; i.e. low in the cerebral ganglion while high in the haemolymph and ventral ganglia by day and *vice versa* at night. The activity of extracts of the corpora cardiaca and corpora allata again remained unchanged.

The above mentioned experiments were discussed in a review by Truman (1971), who concluded that the results had not been confirmed by other authors. The site of the cockroach 'clock' is now thought to be in the optic lobes of the cerebral ganglion in these insects, as indicated by the experiments of Nishiitsutsuji-Uwo and Pittendrigh (1968). Extirpation of the lobes results in loss of rhythmicity, not due to surgical trauma, in the experimental insects. Blackening the eyes or severing the optic nerve has a similar effect, even if the ocelli are intact. As a result of the advanced research nervous control with synaptic transmission is presumed in this case.

The eclosion hormone

The other pattern of circadian rhythm is the one influencing the incidence of eclosion (i.e. adult emergence) and moulting. But, even in this case the control mechanism (the 'clock') seems to be be localized in the cerebral ganglion. In this case, phase setting is accomplished by nervous perception of light by the brain, which, in turn, activates release of the eclosion hormone by neurosecretory cells. The hormone is produced by the brain and reaches the circulation by means of the corpora cardiaca, where it accumulates and is emptied into the haemolymph on the day of eclosion. As well as freeing the adult body from the pupal exuviae, the eclosion hormone is concerned with the switch from pupal to imaginal behaviour, even if the pupal cuticle is removed experimentally at an earlier stage. The eclosion hormone has been found in the silk moth *Antheraea pernyi* and in the tobacco hornworm *Manduca sexta*.

Circadian hormonal timing of larval ecdyses in the same species is to some extent dissimilar. Ecdyses occur at specific times of the day which are instar and species specific and depend upon the photoperiod and temperature in the relevant determination period. Determination occurs by influencing the timing of production of the hormones which initiate moulting. It is possible that AH is responsible for production of the other two metamorphosis hormones and cosequently for the timing of the moulting process.

Neurohormones and hardening of the cuticle

Instantly after ecdysis, the insect cuticle is mostly soft and white. It was shown in a number of insects that the process of harden- ing and darkening the cuticle is also under neurohumoral control from the brain through a neurohormone known as *bursicon*. On the basis of earlier findings by Fraenkel (1935) and Cottrell (1962), bursicon was found to act in adult cyclorrhaphous flies immediately after emergence. It is claimed that bursicon is produced both by the median nsc of the cerebral ganglion and by the compound thoracic ganglion (from which it has been extracted).

It is assumed that it is a polypeptide with a molecular weight of c. 40,000. It has also been shown in *Periplaneta americana*, in *Tenebrio molitor, Galleria mellonella, Oncopeltus fasciatus* and *Schistocerca gregaria*.

The hormone is found in the haemolymph several minutes after ecdysis, reaches its maximum 30 to 60 min. later and disappears completely in a matter of hours. Similar bursicon timing was observed in cockroaches, where it appear to occur in most of the nerve ganglia, but is supposed to be released in the largest amounts from the last abdominal ganglion.

It has been asserted by Fugal and Fraenkel (1969) claim that it also influences melani- zation and continued cuticle formation in the adult fleshfly *Sarcophaga bullata*. Some questions regarding the action of bursicon still remain open, e.g. its relationship to AH and AH and MH action, the hormonal dependencies of cuticle deposition, which start long before bursicon appears, in the preceding instar. The dependence of a single process like cuticle deposition on two quite different hormones (MH, a steroid, and bursicon, a peptide) is likewise not easy to understand. There are further cases in which no similar factor is evidently necessary, e.g. the ecdysis of isolated pupal abdomens in the experiments of Williams (1947, 1948), and other entomologists.

Many research papers have been published on the effect of MH on pupariation in cyclorrhaphous flies. It was suggested that MH is not the only factor responsible for puparium formation, but that a brain neurohormone different from AH also participates. This hypothesis is based on the observation that if blowfly larvae are ligated after the critical period for MH, the anterior part of the body pupariates several hours sooner than the posterior part. If the posterior part is injected with haemolymph from a pupariating larva, or with active brain extract (in insect- Ringer), however, the posterior end can be induced to moult sooner than the anterior end Zdarek and Fraenkel, 1969). The researchers deduced that a specific hormone, which accelerates puparium formation and the inhibition of tanning and potentiates the action of MH, is produced in the pars intercerebralis. This hormone is accepted to be different from both AH and bursicon.

The theory that has been put forward is that other processes of insect ontogenesis might also be under neurosecretory control, such as 'plasticization' (softening) of the cuticle in blood-sucking insects like *Rhodnius prolixus*, which facilitates distension of the abdomen and thereby the intake of larger amounts of blood. Neurosecretory axons of the abdominal nerve supplying the epidermis were found in this connection. Active absorption of the fluid filling the trachea before eclosion from the egg and in connection with larval ecdysis has also been considered as being under hormonal control,and bursicon might be responsible in such a case.

Roussed (1971) studied the Changes in the rate of heart-beat during postembryonic develop- ment were studied in detail by Roussel (1971) in *Locusta migratoria*. It was found to diminish progressively with the approach of meta- morphosis and its course was similar in each instar: it increased at the beginning of the instar, reached the maximum about one-third of the way through the intermoult period and then decreased, attaining a minimum at the end of the second third and then rising again slightly towards ecdysis. It was faster in adult insects and was slightly higher in males than in females. It rose with the temperature up to a given limit (42°C) and was reduced by CO_2 and starvation. It was also affected by the corpora allata and, to a lesser degree, by the prothoracic glands. There was vortually no effect on the implantation of nsc of the pars intercerebralis and of the cc.

Neurohormones and the colour change

In animals, two types of colour variations take place. In one change of colour (or in insects mostly only the change in intensity of colour) results either from the concentration or the dispersion of pigment granules contained in special pigment cells (or in all epidermal cells as in *Carausius morosus*), or from the concentration or expansion of melanophores as in the tracheal air sacs of *Corethra* larvae (*Chaoborus crystallinus*). This type of colour change is a short-term process which is usually cyclic, mainly in a 24-hour period. It results in a rapid adaptation by the insects to the colour or light intensity of the environment. Adaptation of such kind is called physiological colour change. By contrast, the term morphological colour change refers to long-term or irrevocable changes in the colour of the integument. These are produced in the course of ontogenetic development, often as a response by the organism to specific conditions of temperature, humidity, light or other conditions. Such changes have an equally adaptive character. Between the two types of colour variation there are examples of other colour changes. In many case, both physiological and morphological colour changes are regulated by hormones.

The colour variiation are important from the point of view of natural selection. The change may consist in the adoption of protective (cryptic) colouration in response to circadian or other environmental changes, e.g. changes caused by differences in humidity (Joly, 1968), temperature, or it may be related to- thermoregulation, or to display associated with mating and parental behaviour. Mostly, the rhythmicity and intensity of the change is correlated with the important changes in the envorionment.

Physiological colour change

Colour change of the stick-insect epidermis

Comparitively, the cases of physiological colour change occur in insects, are few in contrast to crustaceans where it is a very common feature. Only two cases have so far been described in some details. One of these is the 24-hour change in the intensity of colouration in *Carausius morsus*. This change in colour is resulted from the movement of pigment granules inside the epidermal cells. Four different pigments and found in *Carausius:* time a green and a yellow pigment which are distributed uniformly in the cytoplasm of the epidermal cells, an orange pigment dispersing and concentrating horizontally on a plane with the nucleus, and a dark brown pigment moving in a vertical direction.

In the day time, the dark brown pigment is found at the internal end of each epidermal cell while the orange pigment is concentrated around the neucieus. This explains the light colouration of the insect in day-time. At night, the dark brown pigment moves towards the outer end of the cell and the orange granules spread out evenly, horizontally across the cell. Conseqently the integumental of the insect darkens in colour.

Therefore, it can be clearly stated that the light conditions effect the movement of pigments. If stick-insects are illuminated at night they become higher and, by contrast, stick-insects kept in constant darkness during the day become darker in colour. These changes, however, only become apparent after a while. If the insects are kept in constant darkness the rhythm of colour change takes several weeks to disappear completely. About the same length of time is necessary to reverse the rhythm by keeping the insects in the dark by day and in the light by night. The reactions of the insects to humidity are comparable with their reactions to changes in illumination. In a high relative humidity the colouration becomes darker, and becomes lighter in a low humidity. The colour change is clearly adaptive. Stick-insects sitting on the twigs of a shrub and with a remarkable cryptic body form and behaviour are very much less noticeable to their enemies if they change their colour according to the intensity of the sunlight.

Gierberg (1928) was the first one to suggest the possibility of humoral mechanism for the colour change in *Carausius* and later by Atzler (1930). A detailed study of the problem was carried out by Janda (1934), who showed histologically that there was no connection between the nerve endings and the epidermal cells. In this way the proved that a direct nervous influence on colour change was impossible. In addition, he carried out a series of ligature experiments. When a nymph was ligatured at the middle of the body so as to prevent

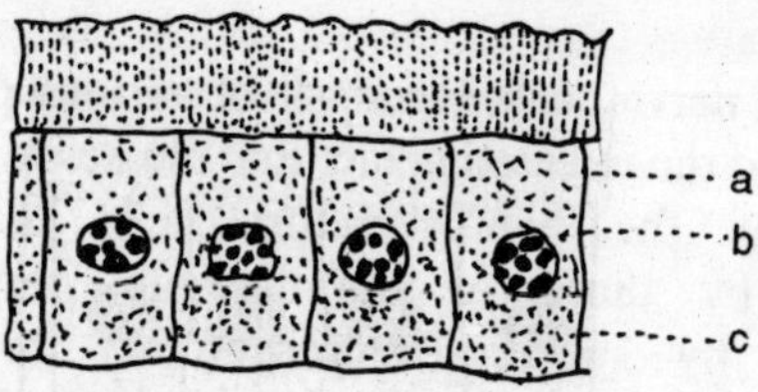

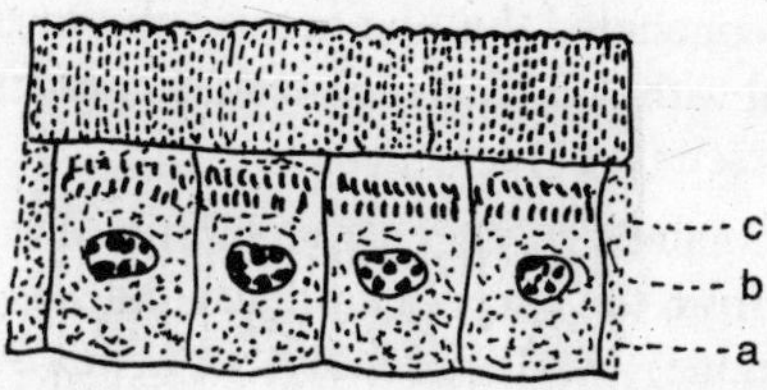

haemolymph passing from the anterior to the posterior part of the body, the diurnal colour change continued only in the anterior part of the body while the posterior part remained constantly light in colour. The ligature caused no. apparent harm to the insect even after many days. When the ligature was loosened, the rhythm of colour change reappeared. After extirpation of the brain the whole body became even light colour. Janda concluded from this that the colour change is initiated by a substance which is released from the brain and transported by the haemolymph. This was confirmed by experiments in which a piece of integument darkened when transplanted from a light-adapted to a dark-adapted insect and, conversely, a piece from a dark-adapted to a light-adalped specimen become lighter in colour.

The assertions of Janda were confirmed and extended by the work of Dupont-Raabe (1951a, b, cf. 1959). She showed that specimens made light in colour by extirpation of the brain become very dark if implanted with the brains of dark-adapted insects. The same effect could be caused by injecting extracts prepared from such brains. The high activity of the extracts was evident from the discovery that even one-twentieth of the amount obtained from a single brain was sufficient to induce darkening. The effect of an extract of the suboesophageae ganglion was equally intensive. A similar, but weaker effect was caused by extracts of the corpora cardiaca and of the ganglia of the ventral nerve cord. Removal of various parts of the brain has demonstrated the interesting fact that the active principal is produced neither by the neurosecretory cells of the pars intercerebralis nor by those of the lateral groups of the protocerebrum, but by specific neurosecretory cells in the tritocerebrum. The existence of these cells was shown by a special histological technique. Raabe (1959) named this substance neurohormone C and concluded that the active substance could not be considered identical with AH. But such results conflict wwith the results acqiured by Gersch (1961)

Raabe (1949-58), obtained another active substance with a similar but weaker effect was from the corpora cardiaca and was designated substance (neurohormone) A. Its identity with the A-substance obtained from the eye stalks of various Crustacea by Carlisle and his co-workers (1955) has been suggested but this is still inconclusive (Carlisle and Knowles, 1959). It does, however, seem to be identical with an active substance obtained form the corpora cardiaca of various insects by Hanstrom (1936, 1940) and M. Thomsen (1943). This latter substance activates the chromatophores of crustaceans. The relationship of these active substances to the C and D neurohormones of Gersch is not yet clear, because of the inadequacy of the technical data; although according to Raabe (1959) there is a possibity ot differentiate amongest them.

A continous Gersch and his collaborators have been able to show that the myotropic substances C and D mentioned above are identical with the substances affecting colour change in *Carausius*.

Extreamely sensitive *in vitro* test was developed, which made possible a detailed analysis of chromatophorotropic activity. Small pieces of *Carausius* integument in insect-Ringer proved to respond readily to neurohormone extracts in far lower concentrations than these which produced a response in whole animals. With this technique it could be shown that neurohormone C_1 causes an intense darkening of the integument whereas D_1 produces less intense   The term 'normal solution' was employed to refer to the quantity of both neurohormones obtained from 10 specimens i.e. from the cerebral ganglion including the corpora cardiaca and corpora allata, from the suboesophageal ganglion, the whole ventral nerve cord and as much haemolymph as possible from each specimen) dissolved in 1 ml of Ringer solution. The term 'Biological unit' is the term which was used to refer to the smallest quantity of either hormone which, dissolved in 1 ml Ringer solution, still produced a darkening of the integument. About 1000 biological units of the two hormones together were found in one specimen of *Periplaneta americana* tested with *Carausices integument. As 50* μg of crystallized neurohormone D_1 were recovered from an extract of 3200 cockroaches, each specimen approximately 0.015 μg.

The chaange in colour is equal in both, in stick-insects and in Mantids. It was shown that light stimuli reach the nervous system solely through the eye. If the eye stalks are cut, the animals do not respond to light changes. Hormonal regulation has not yet been dully elucidated in this case, however.

The experiment called the *in vitro* test was used to prove that neurohormone D_1 is located in the pars intercerebralis (although similar quantities occur also in other parts of the nervous system) whereas C_1 is found in the deuterocerebral and tritocerebral parts of the brain. This would suggest that C_1 and D_1 are identical with the C and A, respectively of Raabe see above). The lack of an affect of extracts from the pars intercerebralis in Raabe's experiments is possibly explained by the fact that at high concentrations its effect is to lighten the integument (Raabe used darkening as a measure of activity). As earlier mentioned, neurohormones C_2 and D_2 are inffecctive, as far as any change in clolur is concerned.

Colour change in Corethra larvae

Yet another illustration of a physiological colour change which has been thoroughly investigated occurs in the larvae of Corethra (*Chaoborus cristallinus*). In these transparent aquatic larvae the colour change is limited to the surface of the air-sacs. There are two pairs of kidney- shaped sacs (in the thorax and in the seventh abdominal segment) of tracheal origin and with a hydrostatic function. All four sacs are covered with cells containing a black pigment and which are capable of considerable distension and contraction. Against a dark background (e.g. the bottom of the pool in which the larva lives) the chromatophores are distended so that they cover the entire surface of the air-sacs. When a larva is transferred to a light background the chromatophores quickly contract, becoming rounded to form only small black spots on the surface of the bladders. Thus the colour change is caused by the movement of whole cells, the melanophores. During the last ten years, it has been proved that that a humoral control is involved.

Systematic research on the mechanism of the colour change in *Chaoborus* was carried out by Gersch (1956; cf. 1960a, b) and his school. A series of experiments, involving the stimulation or removal of individual ganglia in the nerve cord as well as the injection of extracts from various parts of he nervous system; have shown an interaction of at least three neurohormones with partially antagonistic effects. Subsqiemt separation of extract with paper chromatography and electrophoresis has shown that the principal substances concerned are identical with neurohormones C_1 and D_1 and acetylcholine. The effect of these substances on the melanophores corresponds with their effect on the movement of pigment in *Carausius*. Neurohormone C_1 causes expansion of the melanophores and therefore a darkening of the air-sacs. The effect of acetylcholine is very similar, whereas neurohormone D produces a lightening of the bladder colour (cf. Fig.)

Though the changes in physiological colour is a common phenomenon in Malacostraca, in some Cephalopoda and in vertebrates, it has only been observed in a few additional species of insects. In none of these has the physiology yet been studied. Thus it has been discovered in some grasshoppers (Acrididae) including *Kosciuskola tristis*, in the family Oedipodidae, in some mantids and in many other phasmids besides *Carausius*. Undoubtedly even here neurohormones are again involved in the colour change of these species.

Principles affecting colour change

The factors which cause the colour change from both insects and crustaceans were isolated by the above-mentioned methods and were tested on crustacean and insect chromatophores, and muscle contraction rhythms in a detailed comparative study by Gersch, Unger, Fischer and Kapitza (1964). Paper chromatography and paper electrophoresis of various parts of the neurosecretory system of both crustaceans (*Leander adspersus* and *Crangon vulgaris*) and insects (*Periplaneta americana* and *Carausius morosus*) showed the same four different substances in each group: two neurohormones, C_1 and D_1, which already have been found to be of a peptide nature, acetylcholine and serotonin. The substance which was obtained from insects

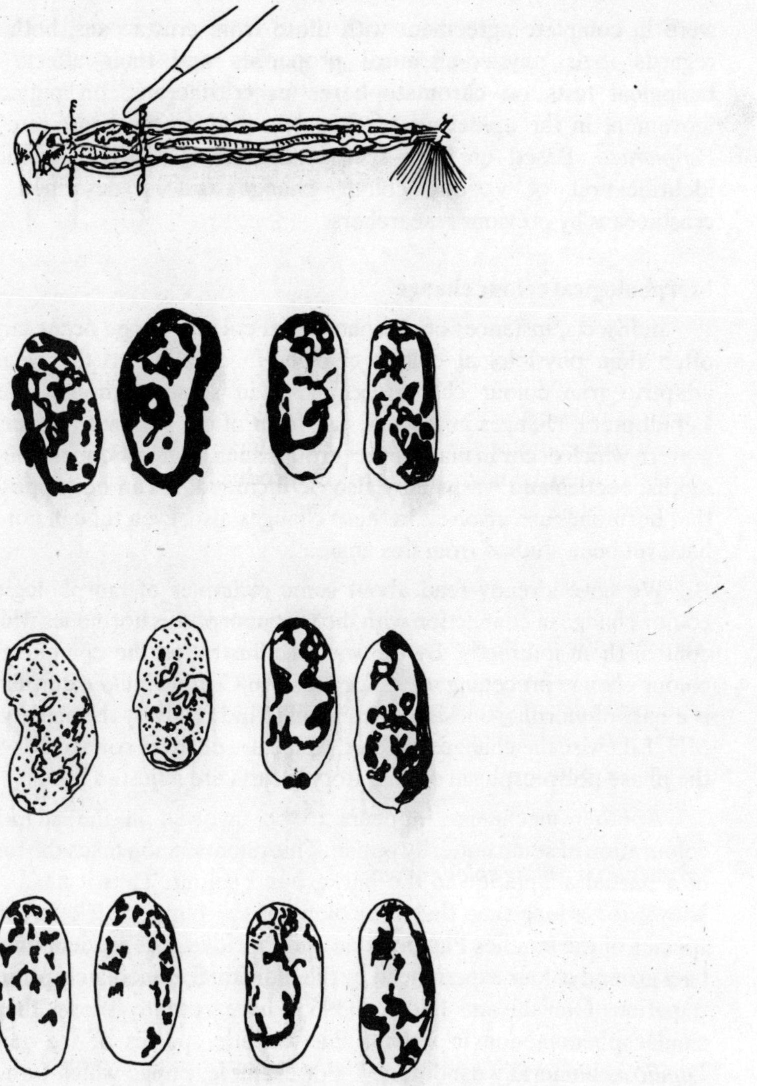

Fig. 3.22. The effect of the neurohormones on the melanophores of the tracheal sacs in *Corethra larvae*. Above-diagram of the larva, into the ligatured thorax of which the hormone is injected, the untreated hind pair of sacs serves as a control (right pair in each series). Upper series – insect – Ringer control; middle series – neurohormone C (left pair of bladders); lower series – neurohormone D. (After Gersch, 1957.)

were in complete agreement with those from crustaceans, both as regards their physico-chemical properties and their effects in biological tests on chromatophores in crustaceans, on pigment movement in the epidermis of *Carausius* and on the heart rate in *Periplaneta.* Based on such theroy suggestions for the chemical identification of various colour change factors described in crustaceans by previous researchers.

Morphological colour change

In insects, instances of morphological colour change occur more often than physiologial colour change. In addition to the clearly adaptive true colour change occurring in some Orthoptera and Lepidoptera, changes in the size and form of the pigment (melanin) pattern which occur in many insect groups such as grasshoppers, bugs, moths, beetles and wasps may also be included. It can be supposed that hormones are involved in these changes also, even though not all have yet been studied from this angle.

We have already read about some examples of morphological colour change in connection with the metamorphosis hormones which control them indirectly. By the way of iillustration the complicated colour change preceding metamorphosis in *Cerura vinula* caterpillars is a part of morphogenesis and controlled indirectly by the activity of MH. Likewise the changes in colur, which are changes connected with the phase polymorphism of migratory locusts are adjusted by JH.

Another mechanism appears to be involved in the adaptive colouration of some butterfly pupae. This phenomenon takes the form of a partial adaptation to the background colour. Thus it has been known for a long time that the colour of the pupal cuticle in some species of the families Papilionidae and Pieridae is dependent on the background colour experienced by the individual immediately prior to pupation. Ohnishi and Hidaka (1955) have recently shown that a similar phenomenon in several East Asiatic species of the genus *Papilio* is humorally conditioned. For example, pupae which formed on young green twigs are also green whereas those which formed on older twigs with a brown bark (on the plant Poncirus trifoliata) are brown.

A set of experiments which involved ligaturing, extirpation of individual ganglia and nerve interruptions with caterpillars before pupation Hidaka (1956) found that colour adaptation was conditioned by optical sensations requiring an undamaged cerebral-prthoracic-suboesophageal ganglionic complex.

Basically the colour of the pupa is green, which might be supplemented by brown depending on the colour of the background. When any of the earlier mentioned ganglia are removed or their connections broken, the brown colour fails to develop even in pupae determined for brown. Both a nervous (optical) impulse and a neurohormonal mechanism are therefore necessary for the appearance of the brown colouration; this is similar to the conditioning of embryonic diapause. Recently Hidaka (1960, 1961) has claimed to have found the source of the pupal brown colour hormone in the prothoracic ganglion of *Papilio xanthus* L. and several other species. In contrast, the brown pupae of several other species of the genera *Nymphalis* and *Malacosoma* which do not show morphological colour changes, do not make similar response when of the prothoracic or other ganglia.

There is hardly doubt that a hormonal mechanism is also responsible for controlling the morphological colour changes in the caterpillars of *Laphygma exigua* and L. *exempts* and *Spodoptera* as well as the seasonal dimorphism in *Araschnia levana* inspite of the fact that there is no experimental evidence of such a mechanism.

Morphological colour changes also occur mostly use adult butterflies and in some caterpillars (of other species, e.g. *Cerura vinula*, see. However all these cases are adaptive (cryptic) in character. For instance, *Pieris, Vanessa* and *Luehdorfia japonica* pupae and caterpillars of certain other species can adapt their colouring to the colour and pattern of the background. Similar adaptation has been observed in many grasshoppers. In some cases, the hormonal mechanism of the colour change has almost been demonstrated in many others, e.g. the spring and summer forms of the adult butterfly *Araschnia levana*, this mechanism can feasibly be presumed. But this is not applicable to the newly described colour change in the *Hercules* Beetle (*Dynastes hercules*) where an endocrine mechanism is impossible

There has beeb some evidence of a connection between JH activity and the shape and size of the ventral black spots on the abdomen of the lygaeid bug *Oncopeltus fasciatus*. Symmetrical black pigment spots develop on the ventral side of the abdomen of adult insects round two foci which correspond in position to the centers of the paired gonads. The differences between male and female in the shape and size of these spots correspond to the differences in the shape and size of the gonads. The number of the spotslessed in insects kept in a high temperature but their change in shape was constant (cf. Fig.)

At the room temperature the spot were largest and practically disappeared at 30°C. Low doses of JH, or supplying the JH by the implantation of corpora allata at the last larval instar, were found to produce the same effect as the high temperature. The influence of increased temperature and the hormone can therefore be explained as a very weak form of progressive metathetly (a delay in development of the imaginal characters produced by the positive influence of other larval parts of the body, including the product of the prothoracic glands). The theory which was suggested that the black spots are controlled by the two sources of a determining substance, which spread out from the parts of the gonads where ripe sex cells first appear. However, castration experiments showed that removal of the gonads at the beginning of the last larval instar made no difference to on the changes in the spots.

Influences on water balance

Van der Kloot (1960), suggested that the close analogy between the insect neurosecretory system of the pars intercerebralis-corpora cardiaca and the vertebrate neurosecretory centers of the hypothalamus-neurohypophysis enable a *priori* that the neurosecretory cells of the insect brain have a role in water metabolism. As illusltrated by many researcchers and proved by various experiments, ot os confirmed, that water balance is influneced by AH.

There are possibilities that suggest that the action of AH on membrane activity, which is manifested in its diuretic effect on water metabolism, may also be the cause of its stimulation of, perhaps, all the glands of the insect organism. Augmentation of the membrane activity of various cellular differentiations, including the nucleus, in other parts of the organism, may likewise explain its general activating character, as in the case of diapause's termination.

Other nse produces elements which occur both in the cerebral ganglion and in other parts of the nervous system, were also shown to affect water balance. Basing their theory on of a histological

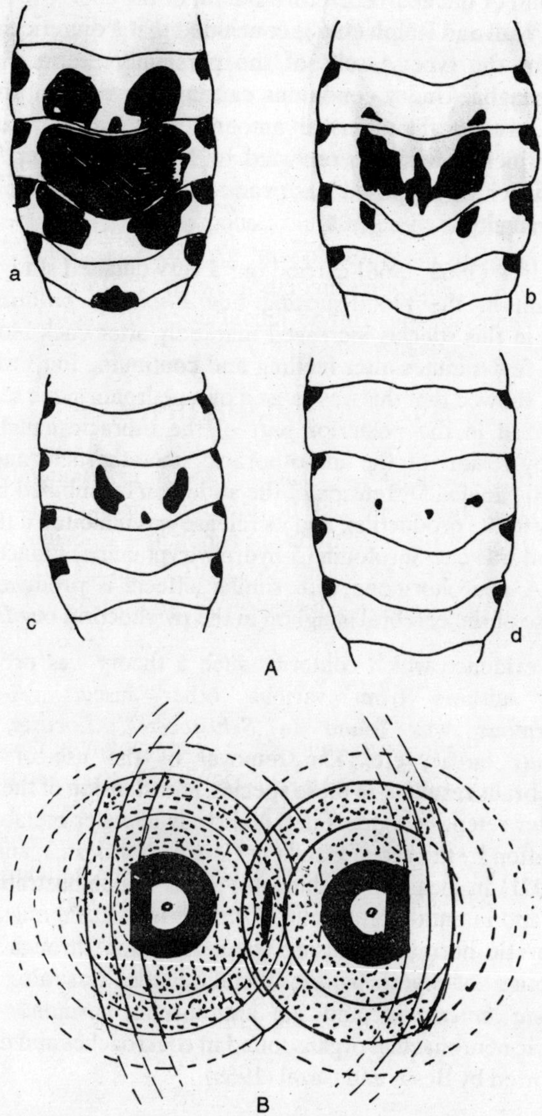

Fig. 3.23. The black pattern of the abdominal sternites in *Oncopeltus fasciatus* females in four different rearing temperatures: (a) room temperature; (b) 25°C; (c) 28°C; (d) 32°C. B, Diagram of the geometrical pattern produced by two factors spreading from two foci with equal and proportionate velocity, inhibited laterally. The areas with equal concentration of the factor are marked in the same way. Compare with the figure above..

investigation of the neurosecretory system of the cockroach *Blaberus giganteus*, Wall and Ralph (1962) concluded that a diuretic factor was released by the type A-cells of the pars intercerebralis and the thoracic ganglia. Under conditions causing dehydration, these cells produced greater than normal amounts of stainable granules. A diuretic principle was also reported by Nunez (1963) in *Rhodnius prolixus,* which he supposed was produced by the nsc if tge scinokex tgiracuc ganglion.

Maddrell (1963, 1966) carried out a very detailed study of water metaboliam in the blood-sucking bug *Rhodnius prolixus*. Water excretion in this species increased markedly after each blood meal, starting a few minutes after feeding and continuing for 3 to 4 hours. Maddrell showed that this was caused by a neurohormone secreted by nsc localized in the posterior part of the thoracic ganglion mass formed by fusion of the mesothoracic, metathoracic and all the abdominal ganglia. Distension of the abdomen by imbibed blood was necessary for its production, and its release was inhibited if the ventral nerve cord was cut. Serotonin (5-hydroxytryptamine) induced similar activity. A neurohormone with similar effects is produced by the medial nsc of the cerebral ganglion in the pyrrhocorid bug *Dysdercus*

The evidence which conforms such a theory was produced by different authors from various other insects. A diuretic neurohormone was found in *Schistocerca, Locusta, Gryllus, Anisotarsus* larvae, etc. The removal of the nsc of the pars intercerebralis resulted, in these species, in distension of the abdomen from water retention made a detailed study of water metabolism and its regulation by the rectal glands in *Carausius morosus*, and Mordue (1970, 1971) in locusts. Mordue succeeded in demonstrating both a diuretic and in antidiuretic neurohormone in *Locusta migratoria*. An anti- diuretic hormone, causing water retention through increased reabsorption of water by the rectal glands, was also found in *Periplaneta americana* and an antidiuretic hormone from the metameric neurohaemal organs found in cockroaches and ctick insect was reported by Besse and Cazal (1968).

Effects on gut proteinase activity

There is a significant unfluence of the neurosecretory products from the brain for normal activity of the gut proteinases in *Calliphora* has been clearly demonstrated. It is doubtlful whether a special hormone is involved or, it is the effect of AH. The probability of the latter is mone than that of the former.

Dadd(1961) illustrates a compasable dependence in *Tenebrio molitor* and by Slama (1964), who studied the effects of removal of the brain nsc and the cc on the haemolymph protein level in *Pyrrhocoris apterus* adults. Hill (1962) and Highnam (1962) made a detailed study of neurohormonal control of protein metabolism in *Schistocerca gregaria*. It was indicated that various factors causing neurosecretion release from the nsc (e.g. hypreactivity, copulation, etc.) resulted in an increase in the haemolymph protein concentration, while conditions inhibiting its release reduced the haemolymph protein concentration. This happened after cauterization of the medial nsc, whereas the reimplantation of active cc into such insects led to a rapid, though brief, increase in the protein level. The protein level was likewise raised by removal of parts of the ovaries. Protein metabolism and the influence of the meta- morphosis hormone was investigated in a series of studies by Janda and Slama (1965), Janda (1970), Janda and Socha (1970), Janda and Sehnal (1970).

Activation of the endocrine glands

The most well-known and most fully analysed effect of neurohormones in insects is the effect of AH on the activity of the prothoracic glands. The findings of Gersch (1961) have produced a new view on this effect. By injecting purified *Periplaneta-americana* brain neurohormone extract into lighted *Calliphora* larvae, as in the *Calliphora* test, simultaneously implanting inactive ring glands, Gersch (1962) was able to induce pupariation. Without simultaneous ring gland implantation this did not occur. Further research led the entomologist and his co-workers to distinguish between neurohormone D and the two AH components. Brauer (1973), in the same laboratory, showed that AH I (the high molecular weight component) strongly stimulated RNA synthesis in the pgl in the second half of the last larval instar of *Periplaneta*. AH II was inactive in this respect. At first, mRNA production was stimulated, while in later phases ribosomal RNA was mainly synthesized. Many researchers had previously demonstrated an increase in RNA synthesis in the pgl of other insects, in the corresponding periods, by means of autoradio- graphy

Not much has been written on the way in which the brain neurohormones act on the corpora allata, or possibly inactivate them, as assumed by some authors, e.g. Engelmann (1965), etc. In principle, however, it does not seem to vary from the mechanism of activation of the pgl.

Effect on the gonads

The most important substance is AH which is necessary for the development of ovaries in *Calliphora* and no doubt also in most other insects, has been shown by E. Thomsen. An increase in the amounts of neurosecretory granules in the neurosecretory cells of the pars intercerebralis in the period shortly before and during egg- laying has been observed in several insect species, e.g. the stick-insects *Carausius morosus* and *Clitumnus Bombyx mori*, *Calotermes flavicollis* and in *Iphita limbata*. Brain hormone is vitally important in the development of the gonads therefor seems to be quite definite.

Yet another interesting connection is that which between the neurosecretory cells and the ovaries has been demonstrated by Scharrer (1955a, b) in the cockroach *Leucophaea maderae*. a striking change in the appear- ance of the type B-cells occurred in adult females which were castrated between the IIIrd and VII instars. In these cells the granules increased in size and number and stained green with paraldehyde fuchsin and Masson-Foot. The neurosecretory material in the other cells remained unchanged. Scharrer concluded that there was a humoral relationship between the type B-cells and the ovaries, and therefore named them castration cells or ovariectomy-cells. However no such effects of castration were observed in males.

The effects of castration were also analysed by Gersch, Fischer, Unger and Koch (1960) in *Periplaneta americana* and *Blatta orientals*. Following castration, neurosecretory cells of type B with granules staining green were also found in other ganglia of the ventral nerve cord. Similar cells aware also found in the ventral nerve cord of *Chaoborus* larvae Undoubtedly, they are generally found in insect. The observed effects are not necessarily a proof of a specific endocrine principle of the ovaries influencing the neurosecretory cells concerned. Probable that the changes that were observed in the 'castration' cells are the result of a falling off in the consumption of their secretion by the ovaries. Therefore the phenomenon is probably similar to the observed hypertrophy of the corpora allata following ovariectomy. In this connection it is very interesting that Noirot (1957) has observed the appearance of new neurosecretory cells in the suboesophageal ganglion of functional reproductives in termites. Another possibility is that the green colouration observed might be connected with an increase in the RNA content of neurons with injured axons, as described by Cahn (1962). But this discovery needs more amount of enquiry.

New data on the hormonal control of metabolism byneurohormones

Fresh information has been obtained by a number of authors in the last three years. They were recently summarized by Gersch (1964). Some of them are discussed in connection with the activation hormone. Among the most important, mention should be made of the finding by Steels (1963) of a diabetogenic factor in the corpora cardiaca of *Periplaneta americana*, reminiscent of that produced by the sinus gland in crustaceans. The injection of corpora cardiaca extracts (even 0.002 of the amount of one pair of corpora cardiaca) resulted in a significant increase in trehalose in the haemolymph. Preaumably that this effect depends on activation of the phosphorylase responsible for the decomposition of glycogen.

The research by Ralph and McCarthy (1964) reveald an increase in the trehalose content in the haemolymph of adult *Periplaneta americana* after injecting them with saline extracts of both corpora cardiaca and brain. They found that the active factor of the corpora cardiaca is produced by the neurosecretory cells of the brain.

Effect on saccharide metabolism

A number of research work on the action of neurohormones affecting the haemolymph sugar concentration has been published. In a study of hyperglycaemic activity in *Periplaneta americana*, Steel (1963) showed that two hormones present in the corpora cardiaca were responsible for raising the haemolymph sugar concentration. Their effect seems to consist in conversion of the enzyme phosphorylase in the fat body from an inactive to an active form, which then catalyses glycogen decomposition, thereby increasing trehalose production. Similar investigations in locusts showed that two hormones released from the corpora cardiaca had the same effect on phosphorylase activity. One (the less active) was produced in the cerebral ganglion and the other was produced by the glandular components of the corpora cardiaca. There was some evidence of interaction. Saccharide metabolism and the influence of various hormones was studied in detail by Nemec (1971), Divakar and Nemec (1973) and Nemec and Rohdendorf (1973).

Effect of neurohormones on electrical activityg

Gersch and Richter (1963) studied about this effect on the nervous system in insects. They showed that extracts of the brain and the ganglia of the ventral nerve cord, like those of the corpora cardiaca, release specific electrical impulses in the phallus nerve of *Periplaneta americana.* The separate application of the neurohormones C_1 *and* D_1 isolated by paper chromatography showed that D_1 possessed strong activity, while C_1 was inactive. The effect of these two substances on the autorhythmic electrical activity of the brain of *Periplaneta americana in situ* and of the isolated mesothoracic ganglion was studied by Strejckova, Servit and Novak (1965). It was found that neurohormone D_1 evoked hyperautorhythmia in both cases, while neurohormone C_1 inhibited the existing electrical activity after first inducing a transient rise.

THE PHYLOGENETIC ORIGIN OF NEUROHORMONES AND NEUROSECRETION

Everyday occurrence of neurosecretion in all multicellular animals (except perhaps Coelenterata) and the striking analogies with insect neurosecretion to be found in the neurosecretory system of the higher animals, leave on doubt of the common principle and phylogenetic origin of this function.

There has been much speculation regarding the phylogenetic relationship between the neurosecretory cell and the nerve cell. Hanstrom (1954) emphasized the basic similarity of the neurosecretory cells and the brain cells. From his detailed comparative research on the anatomy and histology of the nervous system in bothe invertabrates and vertebrates, he suggested the following sequence for the phylogeny of the typical neurosecretory cell: (1) typical neurone; (2) neurone with all the characteristics of a nerve cell but containing a limited number of neurosecretory granules; (3) a neurosecretory cell with axon, dendrites and all other characteristics of a typical nerve cell; (4) a typical neurosecretory cell with axon but without dendrites and the other properties of a nerve cell; and (5) a secretory cell of neurosecretoy origin without any of the characteristics of a nerve cell from the medulla adrenalis of the higher vertebrates.

Another entomologist Clark (1956) based his theory on his investigations of the brain in the polychaete families Nephthiidae and Nereidae. He supposes that secretory activity was phylogenetically the primary function of the neurosecretory cells, and that these were only secondarily included into the mass of the brain. He regards the cerebral organs of the Nemertini, the frontal organs of Phyllopoda (Copepoda) and the X-organs of Malacostraca as evidence which is sufficent to prove his theory.

Gersch (1960), on the basis of his co-worker, Uhde's, work showing the existence of neurosecretory cells in Platyhelminthes (*Dendrocoelum* sp.), agrees with Hanstrom that the nervous system is primary. The neurosecretory cells developed only secondarily from the nerve cells and parallel with the development of the gland cells from the original ectodermal cells. When such a course of action was followed, they gradually acquire the functional importance they now posses in the neuro- secretory system of higher animals.

Significant cofirmation of the secondary character of the neurosecretory cells is their occurrence in very varied parts of the nervous system and in different histological units.

The two contradictory concepts of Hanstrom and Gersch on the one hand and Clark on the other has recently been reconciled by Novak and Gutmann. They have shown that the secretory granules found in the glia cells in the nervous system of cockroaches agree in all their staining properties with the typical neurosecretory granules of the type A-cells. The granules in the glia cells, of course, lack the other physiological features of neurosecretion and do not show any signs of cyclicity and hormonal activity. The authors suggest the term 'gliosecretion' for this particular type of secretory activity. They assume that the gliosecretory granules have the same chemical composition as the carrier substance in neurosecretion but lack the physiologically active neurohormones. Both gliosecretion and neurosecretion are understood to have developed from the secretory activity of the primary ectoderm cells. The same origin is sahied by the so-called trabeculae, a network produced by a particular layer of glia cells round the neuropile of the ganglia, as well as by the perineural lamella, and by the perineural lamella produced by the perineurium cells. The primary secretory product inside the cytoplasm of particular nerve cells then became the carrier substance for the phylogentically later neurohumoral factors (specific to nerve cells) which thus developed into true neurohormones. A detailed cytochemical and electromicroscopic study of gliosecretory granules (gliosomata) has been made by Pipa *et al.*, (1961, 1962).

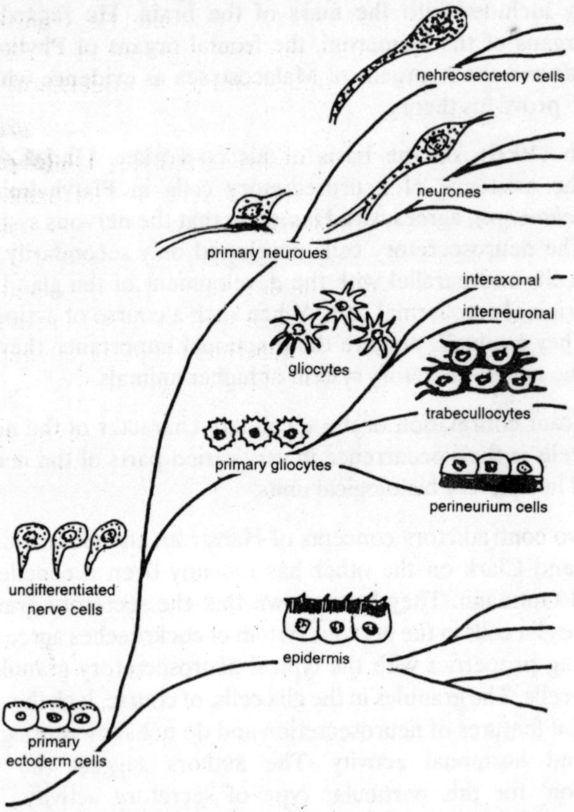

Fig. 3.24. Supposed phylogenetic relationship of the Gomori-positive structures in insects.

One part of the neurosecretory material, the carrier sub- stance, would therefore be phylogenetically earlier than the genera- lized nerve cells, whereas the other, i.e. the neurohormones them selves, would be phylogentically secondary having developed from the neurohumoral factors of the nerve cells as supposed by Hanstrom and Gersch.

4

Pheromones and Allomones

Taste and smell, which can be also termed as chemical sences are much more important to insects than to humans. Insects conntinuously receive with meaningful chemical signal that tell of food, of nest sites, of prey, of predators, or of a suitable mate. Many of these signals produce a rapid and particular behavioral response. The study of the chemical signals that act between organisms constitutes a new discipline called chemical ecology. Much study in this field has resulted into several recent books and innumerable research articles, and is one of the most exciting frontiers in entomology.

Chemical signals include hormones, pheromones, and allomones. ***Hormones*** are internal signals that are produced by an endocrine gland and that are carried through the blood to act on target tissues within the animal. Chemical ecologists study external signals that are secreted into the environment by exocrine glands and that act between different organisms. ***Pheromones*** act between individuals of the same species. Pheromones attracts the sexes, aid in courtship, announce the presence of danger, mark trails and home territories, and control many other intraspecific interactions. E. O. Wilson and W. H. Bossert, of Harvard University, divided pheromones into two general subclasses: (1) ***releasers,*** which act rapidly to produce a behavioral response, and (2) ***primers,*** which act to modify the physiological state. ***Allomones*** act between different species to benefit the producer. Allomones repel predators, confuse prey, and mediate symbiotic interactions.

Insipte of the fact that the terms *pheromone* and *allomone* are of recently coined, our knowledge of chemical signals began in the nineteenth century for pheromones and even earlier for defensive allomones. Yet it is only since 1950 that the analytical power of organic chemistry has developed to the point that identification of the molecules can be accomplished repeatedly from very small samples. With the resurgence of information on the molecules themselves, is now possible to ask precise questions about their impact on behavior and physiology. The molecular structures areinteresting in themselves but the most fascinating aspect of chemical ecology is the delicate intercourse among organisms that is mediated by the chemical signals.

The factor which are necessary for chemical communication are a source of signal molecules, the chemical signal itself, and a receiver. The source is most often an ***exocrine gland***, which opens on the surface of an insect. Glands may be associated with mouthparts, intersegmental membranes, sclerites, legs, and other body parts. Somehow the products of these glands must cross through the impervious insect exoskeleton. Type I secretory cells lie well below the epidermis itself and pass their secre- tions to the outside through

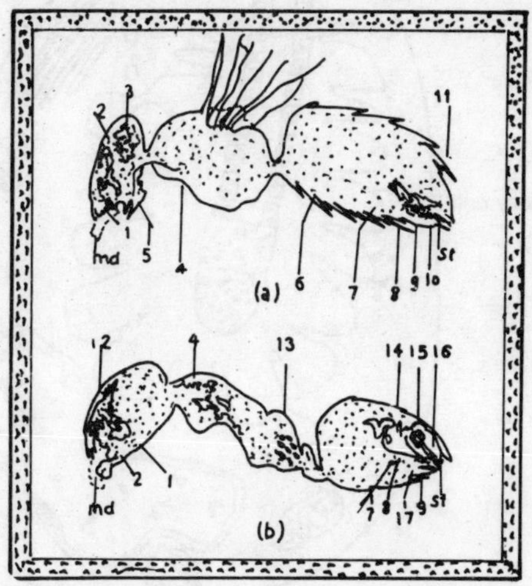

Fig. 4.1. Exocrine gland systems of (a) a honey bee worker and (b) a worker ant (*Iridomyrmex humilis*). Glands of the two species are labeled with the same number if they are homologous. (1) Mandibular glands; (2) hypopharyngeal glands; (3) head labial glands; (4) thoracic labial glands; (5) hypostomal gland; (6) wax glands; (7) poison glands; (8) reservoir of poison gland; (9) Dufour's gland; (10) Koschevnikov's gland; (11) Nasanov's gland; (14) hindgut; (15) anal gland; (16) reservoir of anal gland; (17) Pavan's gland.

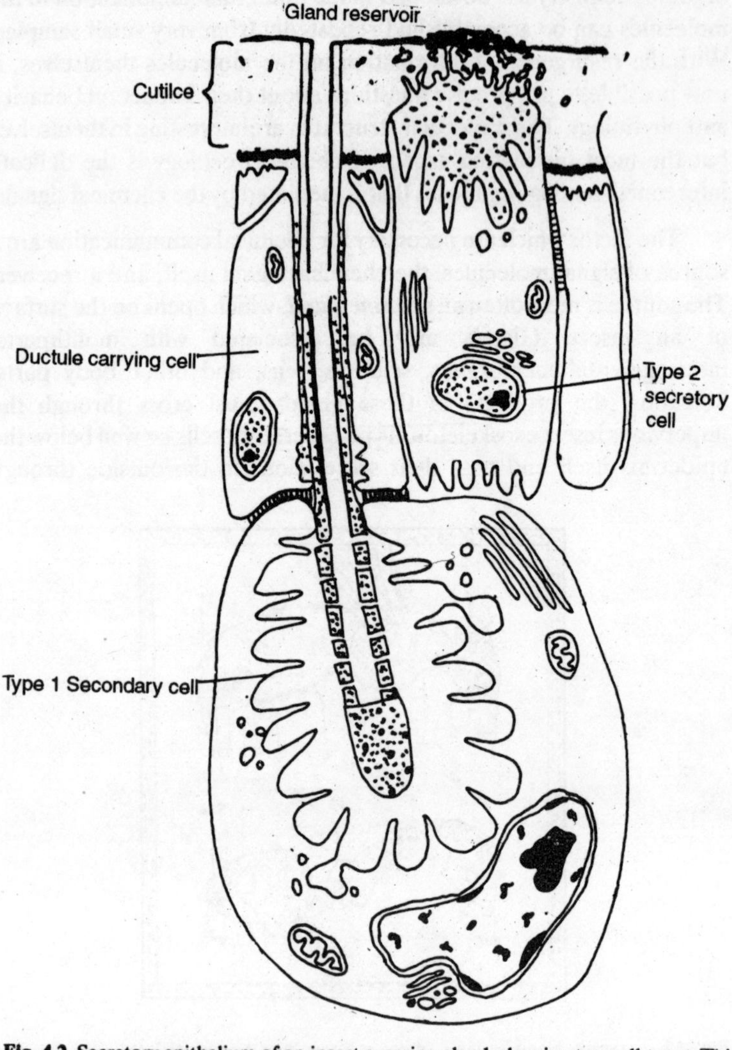

Fig. 4.2. Secretory epithelium of an insect exocrine gland, showing two cell types. This gland regulates the growth of symbiotic fungi that inhabit the cuticular pouch that is the gland reservoir. Type 1 cells pass their products into an enclosed extracellular cavity (cc), within which lies the perforated tip of a cuticular ductule (d1). The ductule (d2) is surrounded by a ductule-carrying cell as it runs to the reservoir. Type 2 cells secrete their products into a pocket of subcuticular space, and the secretions then ooze through the overlying porous cuticle to reach the reservoir.

elongate ductules that are invaginations of the epicuticle. Type II secretory cells have cuticles that are filled with tiny perforations so that the secretion percolates up from the secretory cell to the outside. The molecular signals and their receptors will be discussed below.

Pheromones

From among a noisy jumble of competing chemical stimuli, a pheromone must become comspicuous and convey a clear message. Not all molecules make good pheromones. The signal must be unique and in a terrestrial environment it must be volatile. Glucose and glycogen, for example, are poor candidates on both counts. In theory, the larger the molecule, the greater are the possibilities for a unique

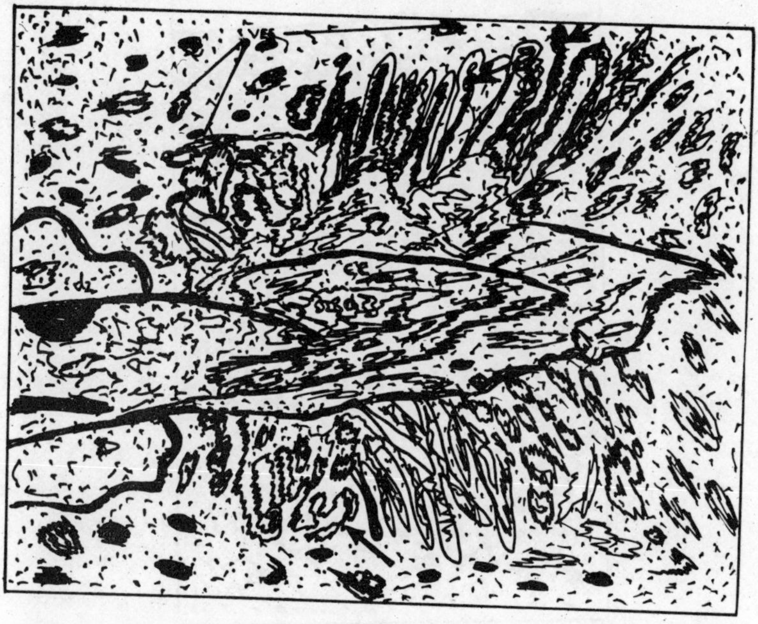

Fig. 4.3. In the fungal transport pouch of a southern pine beetle (Fig. 4.2), a cuticular ductule carries the products of the type 1 secretory cells to the pouch. The secretion is produced in small vesicles (ves), which pass their contents into the central cavity (cc). The wall of the end apparatus (d1) is perforated with fine channels through which the dense secretion flows from the central cavity into the end apparatus and out along the ductule (d2).

structure that conveys a clear message. However, in actuality the requirement for volatility restricts pheromones to molecular skeletons of 20 carbons or less. Most are derived from common biochemicals like fatty acids or amino acids.

In the entire class of Insecta, the molecular structure of most known hormones varies little. In contrast, many thousands of different molecules have been identified as pheromone components. Their diversity among species helps to avoid signal confusion. To increase the information content even more, most pheromones are mixtures of several chemical substances so that in the actual signal several components are mixed togrther.

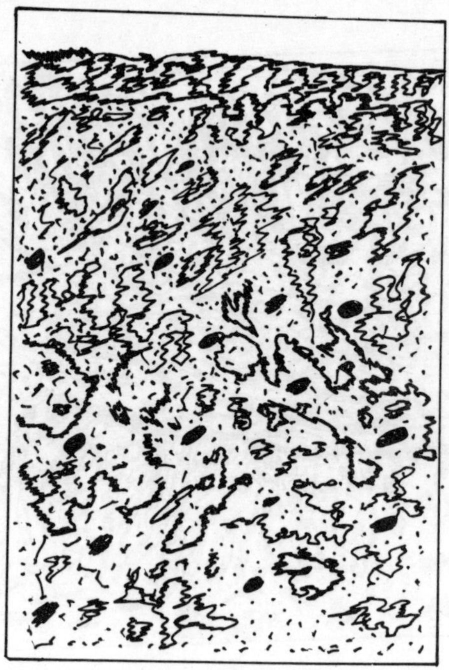

Fig. 4.4. In the fungal transport pouch of a southern pine beetle (Fig. 4.2), the cuticle above a type 2 secretory cell contains many irregular canals through which secretion oozes to reach the space above the cuticle where the fungi are maintained. (Happ, Happ, and Barras, 1971, from *Tissue* and *Cell*, vol. 3, with permission of the publishers, Longman Group Ltd.)

Active Spaces.

pheromone signals occupy space and persist for some time. Many of the special features of the communication systems are related to the expansion of the signal and to its fade-out in space and in time. Imagine an insect sitting in the middle of an empty basketball court and emitting pheromone. Assume that there are no breezes. As the pheromone begins to evaporate from the insect, its concentration rises in the air nearby. As times goes on, the molecules diffuse outward to the edge of the center circle and beyond. There will be a hemispheric cloud of pheromone, with its concentration highest at the source and spreading aroung it.

To begine a chain of reaction the pheromone must be perceived by another insect. In most cases the response is fairly stereotyped, when the pheromone has exceeded a threshold concentration. Far away from the source, pheromone concentrations are extremely low, and no insects respond. As one moves closer to the pheromone source, concentration in the hemispheric cloud reaches threshold. And at lesser distances, concentration is above threshold. Thus there is a zone within which other insects respond and a surrounding environment where they do not. Wilson and Bossert called this zone of response the ***active space.*** In Wilson's words, "The signal is the active space."

The size of an active space is determined by both sender and receiver. When an animal has given off pheromone for a long period, the active space is likely to be at its maximum size. An active space will be larger for a receiver with a low threshold of response than for one with a high threshold. When the emission freuency of the pheromone is high, the active, space will expand faster and achieve a larger volume than when the emission rate is low. By altering the rate of emission, the threshold, and the chemical signal itself, the communication system can be tuned to optimum performance for a particular task. The active space for a sex pheromone may be large, but that for an alarm substance should be quite small.

A basketball court is an artificially created environment, but the lessons learned there about our hypothetical insect can be applied to the natural world. The major differences between the gymnasium and the forest are winds and obstacles. In nature, breezes act on odor clouds. The cloud of pheromone drifts downwind and forms a plume, or aerial odor trail. The plume will rarely be perfectly symmetrical,

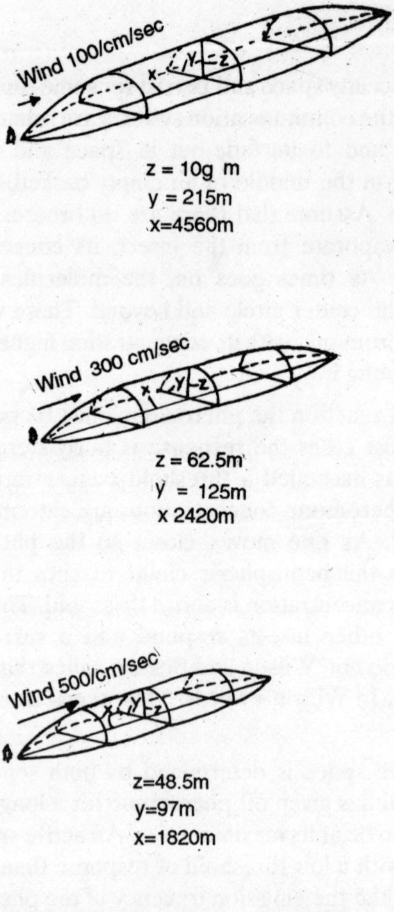

Fig. 4.5. The active space created by a single female moth releasing sex attractant downwind. So long as the female continues to emit pheromone, the space will persist. If wind velocity increases, the active space shrinks.

since breezes increase and decrease unpredictably. The resulting turbulence gives the plume ragged edges, and these edges are further distorted by flowing over hills and into valleys, by going around and through trees, and by changes in temperature and air pressure. Every plume is a little different from all the others. Although there remains much debate about the detailed shapes of these clouds and about the effects of wind and weather, the general concept of an odor plume serves well to give us some clue for the chemical signal.

Sex Pheromones.

Some pheromones attract one sex to the other over long distances, while others act at close range to actival a sequence of the elements in a chain of courtship behaviors. These Pheromones are not produced continuously by all members of a species, but only by animals that are reproductively mature and often only at a particular time of day. Because disruption of reproduction is an appealing strategy for control of insect pests, a large number of species have been studied in university laboratories and agricultural research stations throughout the world. The research began as a descriptive search for the molecules, and as more and more have been identified, the field is shifting toward precise study ofhow pheromones function.

During the period around 1960, it was thought that one sex of each species might produce a single, unique substance that automatically triggered a rigid sequence of stereotyped behavior in the opposite sex. That view of pheromones was attractive because of its simplicity. However, natural selection has left much more interesting and complex phenomena for the entomologist to analyze. Most pheromones are complex medleys of scents, and these odor blends have many subtle shades of meaning. Rather than list all the molecules, or even a representative few hundred, we shall study in detal the biology and chemistry of the sex pheromones which affect the reproduction of two groups of insects: moths and queen butterflies.

Moth Attractants.

The French naturalist J. H. Fiber a nineth century entomologist demonstrated that male peacock moths are attracted to a paper that previously held a female but pay no attention to a female under a glass bell jar. He correctly concluded that a residual sex scent on the paper was responsible. But it was not until 1959 that Adolf Butenandt, of the University of Munich, Germany, became the first to determine the chemical structure of a pheromone. That was the culmination of a 30 year effort with *Bombyx mori*, the oriental silkworm moth, which has been cultured for many centuries in Japan and China.

As soon as a female *Bombyx* moth comes out from her pupal cocoon, nearby males become highly excited and will copulate immediately with her if given the opportunity. Males do not even need to see the female. They exhibit characteristic excitement (wing fluttering, flying upwind, abdominal movements, attempting copulation) when exposed to female scent. generally, the scent is emanated from two glands that open in the intersegmental membrane at the tip of the female abdomen. Butenandt and his colleagues painstakingly snipped off the tips of the abdomens of 500,000 virgin female *Bombyx*. These tips were then extracted with ethanol-ether, and the active components were carefully purified from the crude mixture and then characterized by brilliant microchemical techniques that were devised by Butenandt for this project. He obtained about 12 mg of pure attractant which is termed as bombykol and identified it as a 16-carbon alcohol.

Male *Bombyx* are extremely responsive to female scent, and the male fluttering response is used as a bioassay to compare different concentrations of pheromone. For each test a sample of dilute pheromone is placed on a filter paper cartridge and blown by an air-stream to resting males. A typical dose-response curve for pure bombykol is shown in Fig. 4.8. If carefully examined such curves may seem dry an abstract but they neatly summarize the data and show how they fall into a pattern. Below 10^{-6} μg the response is no different from background activity. Between 10^{-6} μg and 10^{-4} μg there is an abrupt rise as an increasing percentage of males respond. Above 10^{-4} μg, nearly all the males respond.

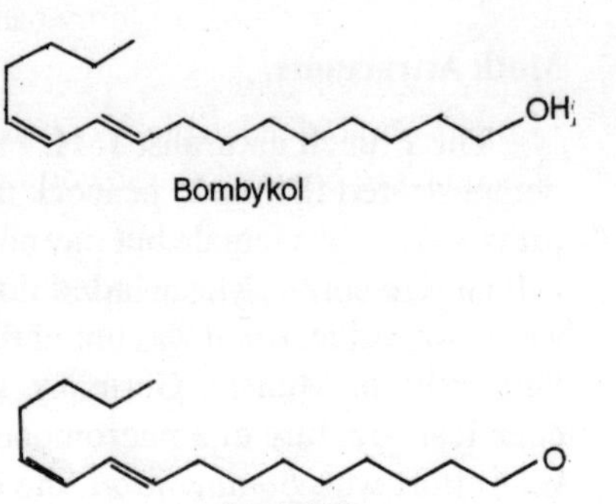

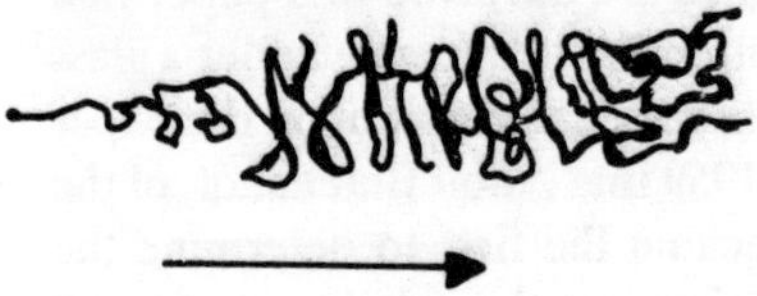

Fig. 4.6. The irregular filamentous structure of an ador plume formed by continuous emission of a pheromone into air moving in the direction of the arrow.

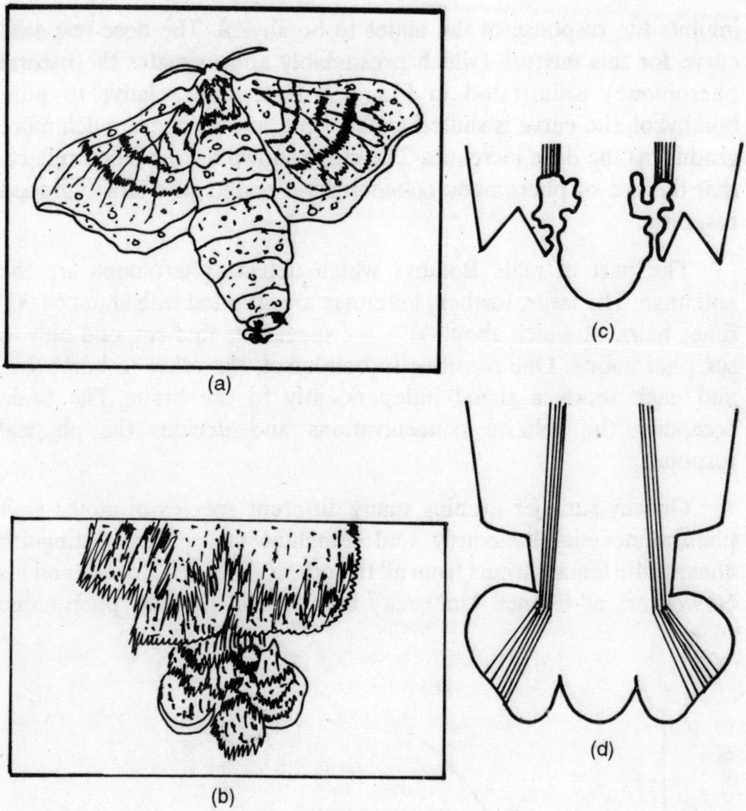

Fig. 4.7. (a and c) The gland of the female silkworm moth that secretes the sex pheromone. The secretory cells open into a pouch formed by the infolded intersegmental membrane. When the female is "calling" (emitting pheromone), the pouch is folded out so that the pheromone can evaporate, as shown in the lower figures (b and d).

For a considerable period of time, it was wrongly presumed that bombykol is the only component emitted by female *Bombyx*. But in 1978 K. E. Kaissling and his colleagues at the Max Planck Institute in Sewiesen, Germany, found that the female also produces the corresponding aldehyde in the ratio (bombykol:bombykol) of 10 : 1. When males are stimulated with pure bombykol at moderate doses, there is no behavioral effect. But when the alcohol and aldehyde are blended together as in the natural pheromone, bombykol partially inhibits the response of the males to bombykol. The dose-response curve for this mixture (which presumably approximates the natural pheromone) isillustrated in Fig. 4.9. Note that relative to pure bombykol, the curve is shifted to the right, and its rise is much more gradual as the dose increases. The curve shown for the blend reflects that the use of pheromone concentration results in a more graded response.

The part of male Bombyx which detects pheromone are the antennae. The large, feathery antennae are studded with about 64,000 sense hairs, of which about 80% are specialists that respond only to sex pheromone. One sensitive to bombykol, the other to bombykal, and each sends a signal independently to the brain. The brain computes the relative concentrations and decides the physical respones.

On any summer evening many different species of moths emit pheromones simultaneously, and somehow males must distinguish conspecific female scents from all the others. Wendell Roelofs and his co-workers at Cornell University have shown that the pheromone

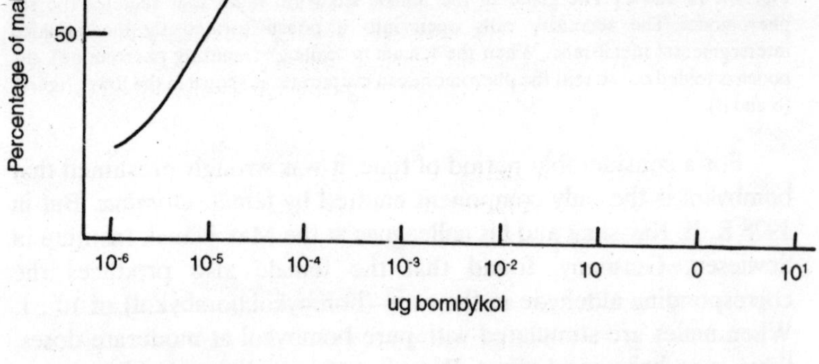

Fig. 4.8. Fluttering response of male silkworm moths when exposed to increasing concentrations of bombykol.

blends emitted by some species are carefullyy timed. For example, two species of the genus *Archips* (leafroller moths) share the same four components in their pheromones. But the two species differ in the relative proportions of the components-60:40:4:200 for A. *argyrospilus* and 90:10:1:200 for A. *mortuanus*. Males of each species have specialist receptors for each of these molecules. Their sensilla have at least four neurons, and apparently each neuron recognizes one of the components. Signals from these sense cells are sent in parallel to the brain, where the relative proportion of the components must be computed. In this way males are able to discern the partilcular type of scent emitted by females of their own species.

Now the question arise that how well can moths discriminate between a signal, such as bombykol, and other molecules of similar size and with similar properties? The fluttering response can be used to compare similar molecules, such as geometric isomers of bombykol. Bombykol has two double bonds and thus four geometric isomers, Butenandt synthe- sized all four isomers, and experiments showed that the threshold for bombykol itself is 100 to 1000 times lower than for any other isomer. Most of the experimemts have resulted into the belief that precise molecular shape is recognized by pheromone receptors.

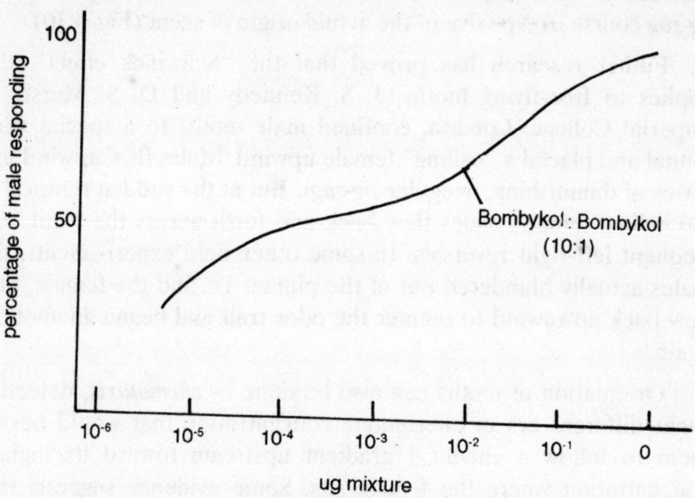

Fig. 4.9. Fluttering response of male silkworm moths when exposed to increasing concentrations of a bombykol:bombykal mixture (10:1).

Another question whifh arised is about the number of male each female can attract. The pheromone gland of each female *Bombyx* contains about 164 ng of bombykol, in theory enough molecules to excite 10^{12} males if each received a threshold dose. But such calculations ignore the impact of bombykal in the blend, and even if so adjusted, they would not be biologically meaningful. The important advertising signal is the active space. Production of large amounts of pheromone is largely an adaptation to maximize the volume of active space. By spreading wide the area of active space, the female increases the likelihood of attracting males.

As soon as attracted by pheromone within the active space, a male requires orientation cues in order to find the female. For long-range orientation, he appears to use mostly ***anemotaxis***-movement based on the direction of wind.

The upurind movement of insects takes them toward the source of wind- borne scent. It is difficult to observe sustained oriented movement in free-flying moths, but Ilse Schwinck, of the University of Munich, Germany, found a clever alternative system. She used a flightless *Bombyx* mutant. When excited by female pheromone, flightless males wave their useless wings and try to run to the female. Schwinck watched the males as they ran along a table in the laboratory. When exposed to female scent, males ran upwind in a *zig-zag* course irrespe tive of the actual origin of scent (Fig. 4.10).

Futher research has proved that the "Schwinck effect" also applies to free-flying moths. J. S. Kennedy and D. S. Marsh, of Imperial College, London, confined male moths to a special wind tunnel and placed a "calling" female upwind. Males flew upwind in a series of diminishing, irregular *zig-zags.* But at the sudden removal of the calling female, males flew back and forth across the wind with frequent left-right reversals. In some other field experi- ments, the males actually blundered out of the plume. To find the female, they flew back downwind to reenter the odor trail and began anemotaxis again.

Orientation of moths can also be done by ***chemotaxis***, detecting slight differen- ces in pheromone concentration that would permit them to follow a chemical gradient upstream toward its highest concentration where the female sits. Some evidence suggests that moths can detect and follow chemical gradients in still air, but it has yet to be shown that they can tell upstream from downstream.

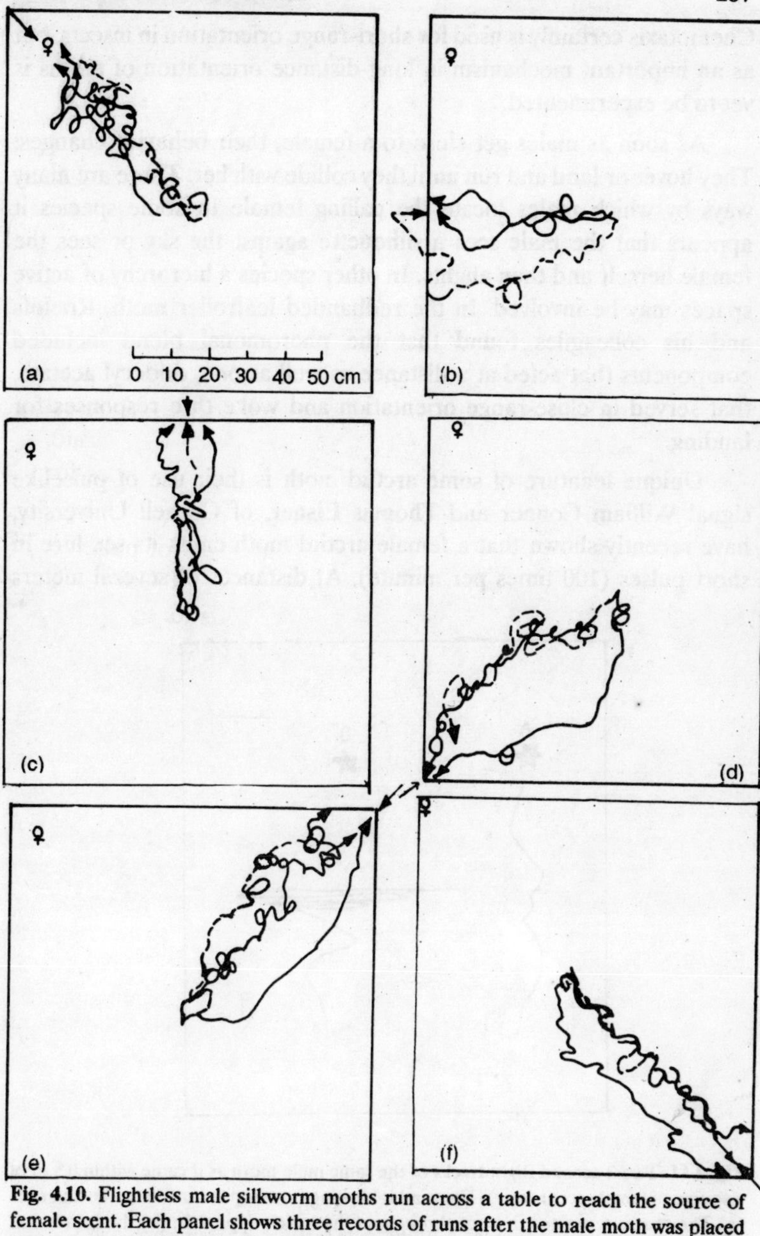

Fig. 4.10. Flightless male silkworm moths run across a table to reach the source of female scent. Each panel shows three records of runs after the male moth was placed in the center of the table. The female is always placed at the upper left corner. The wind direction, indicated by the small arrows, was systematically moved around the table. It is clear that males mc ›e

Chemotaxis certainly is used for short-range orientation in insects, but as an important mechanism in long-distance orientation of moths is yet to be experimented.

As soon as males get close to a female, their behavior changes. They hover or land and run until they collide with her. There are many ways by which males locate the calling female In some species it appears that the male sees a silhouette against the sky or sees the female herself and then alights. In other species a hierarchy of active spaces may be involved. In the redbanded leafroller moth, Roelofs and his colleagues found that the pheromonal blend included components that acted at a distance as well as one, dodecyl acetate, that served in close-range orientation and woke thre responses for landing.

Unique feaature of some arctiid moth is their use of pulselike signal William Conner and Thomas Eisner, of Cornell University, have recently shown that a female arctiid moth emits its sex lure in short pulses (100 times per minute). At distances of several meters

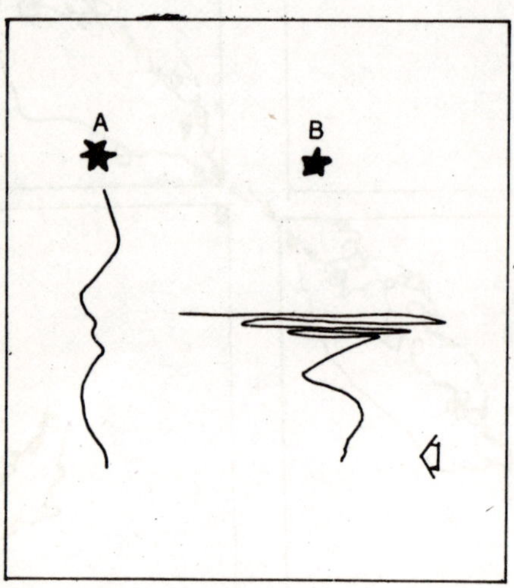

Fig. 4.11. Two 3-second flight tracks of the same male moth as it came within 0.5 m of a source of female scent (star). The wind is blowing from top to bottom in the diagram. (a) The source remained in place throughout the male's approach; (b) the source was removed as the male reached the level of the arrowhead so that the male reached unscented air as he flew 0.5 m upwind.

from the source, the pulses would probably be smeared together by turbulence, but close in, they remain discrete. When males are but a few centimeters downwind, their antennal receptors show corresponding pulsations in electrical potential. It is possible that when the continuous signal transforms into an on/off pattern of pulses, males switch from upwind anemotaxis to the rsearch in shorter range

Queen Butterfly.

Visual signal attract males to females in most diurnal butterflies, such as the queen (*Danaus gilippus berenice*). The courtship sequence, a chain of signals between male and female, was described in detail by Lincoln and Jane Van Zandt Brower while they were at Amherst College in Massachusett.

Male queen butterflies have a pair of large organs, called ***hair pencils***, which are really tufts of fine hairlike processes of cuticle. As soon as a male queen sees a female in flight, he quickly overtakes her. As he passes over her dorsal surface, he extrudes the large hair pencils from the tip of his abdomen and bobs up and down over the female's head and antennae. As demonstrated by Thomas Pliske and Thomas Eisner, of Cornell University, the hairs are covered with dustlike

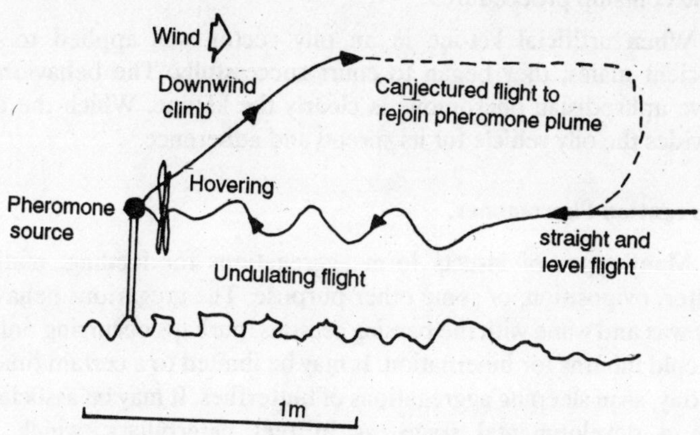

Fig. 4.12. Male moths approaching a pheromone source often hover near it before alighting. If they accidentally leave the odor trail, the males fly downwind and start over.

particles that are brushed onto the antennae of the female. These numerous particles are actually microscopic spherical bits of epicuticle that were formed during pupal development. The particles clump together and stick to the hairs and to the antennae because they are coated with an oil film. The oil contains the real signal, and the dust is just a means for its transfer. The oil is apparently produced by a secretory cell at the base of each hair. Jerrold Meinwald and his team at Cornell identified two major components in the secretion: a ketone and a viscous terpenoid diol.

As a result of to the hair penciling, the female lands. The male continues to hair pencil vigorously, and the female folds her wings. The male then alights and attempts copulation. Pliske and Eisner deprived males of their hair pencils and found that, although they courted actively, the courtship very rarely led to copulation. Males raised indoors on cuttings of the major food plant failed to mate successfully, although males raised outdoors, mostly on the same intact food plant, courted successfully. Comparison of these two male groups showed that the indoor males had the diol but lacked the ketone, whereas the outdoor males had both components. The outdoor males obtained the chemical precursor of the ketone from another plant on which they spend very little time and of which they eat little. Therefore it is clear that the intake of ketone by males is vital in the coutship proceedure.

When artificial ketone in an oily vector was applied to the deficient males, they began to court successfully. The behaviorally active aphrodisiac pheromone is clearly the ketone. Which the diol provides the oily vehicle for its spread and adherence.

Aggregation Pheromones.

Many types of insects form aggregations for feeding, mating, shelter, oviposition, or some other purpose. The gregarious behavior may wax and wane with the passing seasons, perhaps occurring only in the cold months for hibernation. It may be limited to a certain time of the day, as in sleeping aggregations of butterflies. It may be associated with a developmental stage, as in tent caterpillars, which are gregarious as larvae but not as adults. It is generally understood as opportunistic behavior, as when lycid beetles form large aggregations to feed on sweet clover and to mate. Once the temporary or restricted resource is exhausted, the aggregation breaks up. The majority of insect group form in response to pheromones. The most intensively studied species are bark beetles and ambrosia beetles, both members of the family Scolytidae.

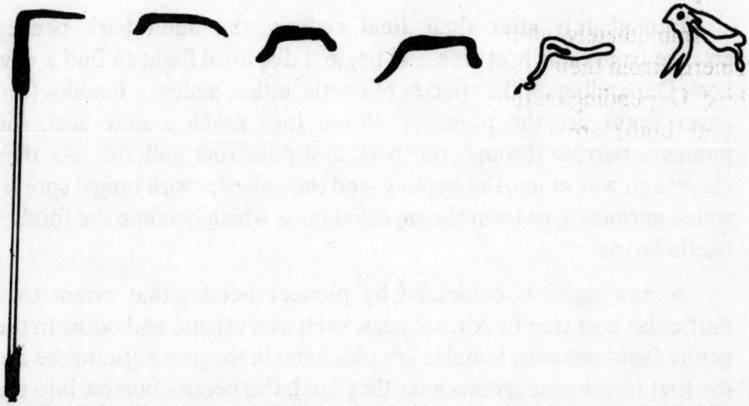

Fig. 4.13. Simulation of a temporally patterned chemical signal. Air laden with a visible marker (titanium tetrachloride) is pulsed at a rate of 2 puffs per second from the glass capillary shown at the left. The pulses remain discrete over a distance of 60 cm.

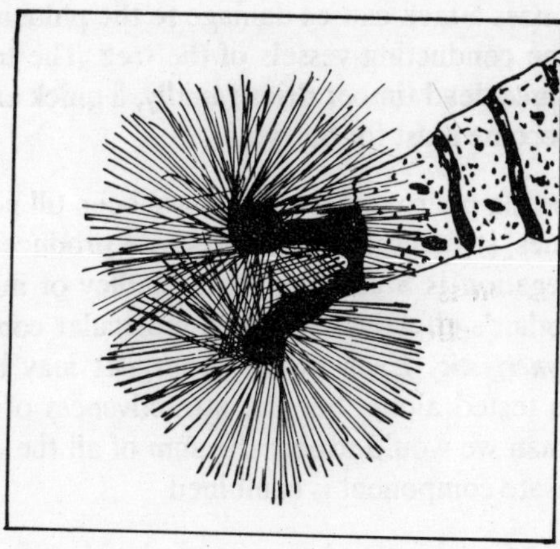

Fig. 4.14. Hair pencils on the abdominal tip of a male queen butterfly.

Bark beetles and ambrosia beetles join together for feeding, mating, and oviposition. Because the beetles and the fungi that they carry from tree to tree do tremendous damage to pines of our forests and to the American elm, he aggregation behavior has been exhaustively studied. Bark beetles' pheromones explain the complexity and natural history of a chemical signal system.

Immediately after their final ecdysis, the adult bark beetles emerge from their host tree and begin a dispersal flight to find a new host. Depending on the species of beetle, either males or females (but never both) are the pioneers. When they reach a new host, the pioneers burrow through the bark and construct galleries. As they chew their way along, the beetles seed the galleries with fungal spores, which germinate to form the mycelial mass which become the food of beetle larvae.

A new tree is colonized by pioneer beetles that orient to a particular host tree by several cues, such as its shape and odor. In the genus *Dendroctonus*, females are pioneers; in the genus *Ips*, males are the first to colonize. As soon as they land, the beetles burrow into the bark. Healthy trees produce a flow of resin or pitch that may eject the burrowing beetle from its partially completed hole or may drown it. But when a few pioneers manage to complete the entrance tunnel and to establish themselves, they release aggregation pheromone. More beetles are attracted, and the balance begins to shift toward the intruders. The mass attack causes damage to the phloem, and fungal hyphae block the conducting vessels of the tree. The tree gradually begins to die. Since dead timber dries rapidly, a quick explaitation of the resources become must for beetles.

The bark beetle pheromones which have been till now identified are monoterpenes, structurally similar to those produced by the host tree. The aggregation is a response to a medley of molecules that enhance each other's effectiveness. Such molecular components are described as ***synergistic***. Each of the synergists may be somewhat attractive when tested alone, but the attractiveness of the blend is much greater than we would expect the sum of all the contributions from each separate component is combined.

The very first bark beetle pheromone to be identified was that of *Ips confusus*, the California five-spined ips. As the beetles chew their way into ponderosa pine trees, the wood passes through their gut and

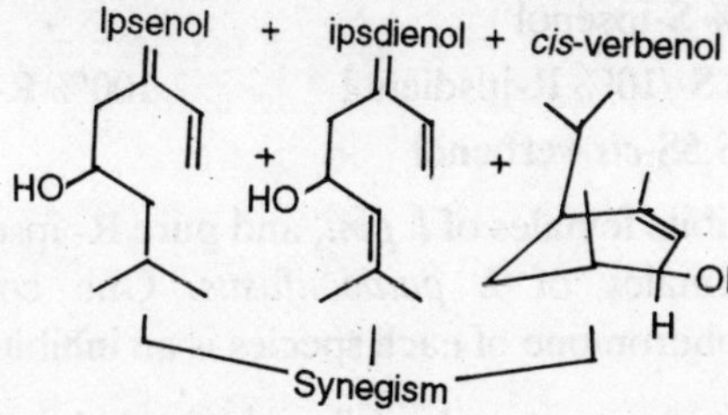

is deposited pellets called frass. David Wood, an entomologist at the University of California, Berkeley, and Robert Silverstein, an organic chemist now at Syracuse University, isolated three terpene alcohols, ipsensol ipsdienol, and *cis*-verbenol, from the frass of attacking male engraver beetles. In a laboratory bioassy with walking female beetles, ipsenol alone was excitatory as a single component, and it was synergized by mixing with one or both of the others. In field tests only the combi- nation of all three synthetic compounds proved attractive. When the synthetic mixture was later compared with "natural" scent from maleinfested pines, that natural sent was much more effective than a synthetic blend of ipsenol, ipsdienol, and *cis*-verbenol. Clearly the natural pheromone contains components which are till now unidentified.

Near the male-infested tree, a female runs about, searching for entrances to a male burrow. Outside its entrance, she stridulates by rubbing the serrated edge of her elytra against a scraper on her abdomen. In response to the female chirps, the male may allow her to enter his burrow; mating follows. One male will allow only three females to enter the nuptial chamber, and, after mating, each female constructs an egg gallery in the phloem and outer xylem. When more females stridulate or try to intrude the burrow, the male uses his bodyto blok the intrance.

Many species of *Ips* have comman geographic distributions. Both *Ips paraconfusus* and *Ips pini* attack ponderosa pine in California, but the two species rarely attack exactly the same tree. Martin Birch and his colleagues at the University of California, Davis, have shown that resource competition is minimized because each species responds only to its own attractant blend and because each blend includes an inhibitory scent for the other species. The pheromones emitted by males is composed:

Ips paraconfusus	*Ips pini*
100% S-ipsenol	
90% S-/10% R-ipsdienol	100% R-ipsdienol
IS,4S,5S-*cis*-verbenol	

S-ipsenol inhibits females of *I. pini*, and pure R- ipsdienol inhibits the attack by females of *I. paraconfusus*. One component of the aggregation pheromone of each species is an inhibitory allomone.

As it has been stated earlier, in bark beetles of the genus *Dendroctonus* it is the females that are pioneers. This genus includes many serious pests of coniferous forests, such as the southern pine beetle and the western pine beetle. From the frass of the latter species, Wood and Silverstein isolated and identified the unusual terpene *exo*-brevicomin. Pure *exo*-brevicomin is quite attractive to males, but many more males were attracted to synthetic mixture of *exo*-brevicominand myrcene, a terpene produced by wounding the host ponderosa pine. After responding to these substances, males at a tree laden with female and gove off a pheromone of their own, frontalin, which attracts other females. Like *exo*-brevicomin, frontalin is synergized by myrcene. The combination of these substances is cause for the mass attack of both sexes. After mating, both sexes begin to release two additional pheromones, *trans*-verbenol and verbenone, which stops the arrival of other beetles.

Inhibitory pheromones are one device to control density so that the tree is not overcrowded with larvae. Another device to control density is high attractant concentration, as demonstrated in trapping experiments. Beetles are readily caught in artificial traps baited with a mixture of brevicomin, frontalin, and myrcene. However, when pheromone concentration gets extremely high, the catch in the trap declines. At very high concentrations of pheromone the incoming beetles trun away from the baited tree and shift their attention to the nearby trees. Consequently a cluster of infected trees is produced, which becomes a "hot spot" of brown foliage in the forest.

It is remarkable that the of beetle receptors to recognize the differences between the pheromone components and similar molecules. Geometric isomers are easily discriminated; thus *trans*-verbenol is an effective inhibitor for attacking western pine beetles, while *cis*-verbenol is not.

It is yer unknown that how the chemicals in bark beetle aggregation pheromones are manufactured. No discrete exocrine glands seem to be responsible. The aggregation scent is not produced if the host terpenes are absent, and many of the pheromones seem to be oxidized derivatives of the host terpenes.

For instance, male *Ips paraconfusus* convert myrcene to ipsenol and ipsdienol. David Wood and his associate, John Byers, have proved that the oxidation of myrcene to ipsenol and ipsdienol is inhibited by the antibiotic streptomycin. The suggest that symbiotic bacteria, res- tricted to the male, are responsible for myrcene oxidation. If bacteria play such a role, the beetle depends on the tree for the starting material and the bacteria for the metabolic conversion in order to produce an insect aggregation pheromone.

Initialting an aggregation carries a risk. Pradaceous carabid beetles, which feed on the bark beetles, are also attracted by the bark beetle aggregation pheromones. When the carabid arrives at the infested tree, it feeds on the beetle which emits the signals.

Alarm Pheromones.

The major disadvantage of such mass gathering of insects is that a predator has an easier time finding prey. Upon approach of the predator, the group may disperse or it may invoke a colonial defence response . In some hemipterans and in social insects, an alarm pheromone resutls in the similar behavioural pattern.

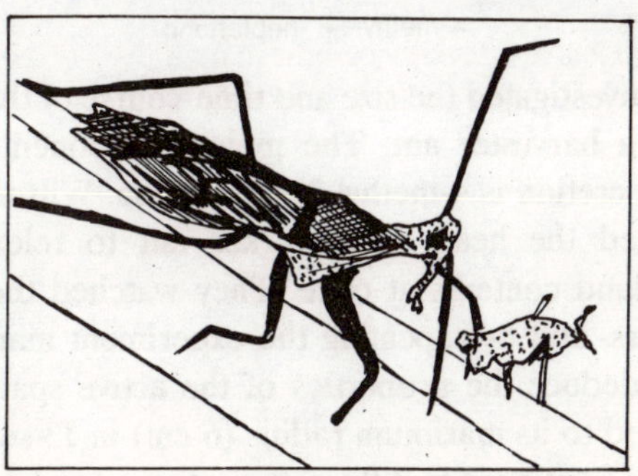

Fig. 4.15. After an attack by a predatory bug (Hemiptera, Nabidae), an aphid secretes a droplet of fluid that contains the alarm pheromone.

Aphids gather in large number to feed. When a predator attacks, the victim secretes a droplet of fluid from a large gland opening called a *cornicle* (Fig. 4.15). Within that fluid is an alarm pheromone, of which the most common is farnesene. Nearby aphids respond by falling to the ground, walking away, or leaping from the plant.

Alarm Pheromones of Ants.

Anyone who has stepped on an ant nest has seen the outrush of defenders, which open their mandibles and swarm over the attacking foot. Such alarm behavior is caused by a pheromone emitted by the mandibular glands and by glands associated with the sting.

It has been predicted by Wilson and Bossert that an alarm signal should be localized, blatant, and brief. The pheromone should spread rapidly to coordinate colonial defense and should fade out rapidly to avoid prolonged false alarms. For the site of the disturbance to be easily located, the active space should remain small. To prove there

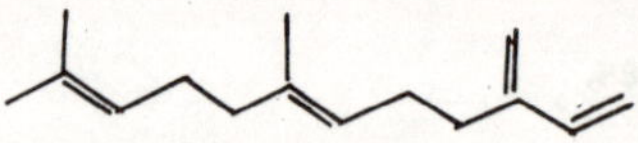

β- farnessane

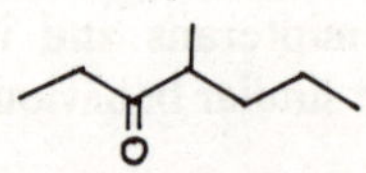

4-methyl-3- heptanone

findings they investigated the size and time-course of the active space for alarm in a harvester ant. The major component of this ant's Mandibular secretion is 4-methyl-3- heptanone. Wilson and Bossert quickly crushed the head of a worker ant to release all of its mandibular gland contents at once. They watched the response of nearby workers, and by repeating the experiment many times, they were able to deduce the properties of the active space. The active space expanded to its maximum radius (6 cm) in 13 seconds, and as the pheromone continued to diffuse, the active zone shrank to nothing by 35 seconds. In a natural situation with a serious invasion of the nest, other workers, excited by the first secretion, would have encountered intruders and added their own pheromone. This added pheromone would amplify and prolong the signal and expand the defensive frenzy. The extent of the group response decide to the severity of the attack.

Alarm pheromones are very volatile, and most are of low molecular weight, containing 12 carbons or fewer. Those from the mandibular glands usually have a keto-or aldehyde group, while those from Dufour's gland (near the sting) are most often hydrocarbons. Many are mildly toxic and distasteful; thus they function as alarm defencive devices.

In case of the african weaver ants construct an elaborate nest by sewing leaves to one another with thread from the silk glands of their own larvae. Both the nest itself and the territory surrounding it are vigorously defended by the major workers. Any insect that happens to wander into that territory is rapidly attacked and often becomes prey for the ants. The defensive behavior consists of four stages. (1) alerting, in which ants raise their heads and open their jaws; (2) attraction toward the disturbance; (3) stopping near the source of the disturbance; and (4) biting and holding for several minutes.

The distress signal follows the expulsion of the fragrant but musty ofor from the mandibular glands of a major worker. J. W. S. Bradshaw, R. Baker, and P. E. Howse of the university of Southhampton, England, have shown that at least 33 volatile components are released by crushing the head of a major worker. Four of these seem to trigger specific aspects of the alarm response. These molecules differ in their volatility from hexanal (most volatile) to 2-butyl-2-octenal (least volatile). Their defense mechanism can be easily apprehended if we imagine that an excited ant deposits a droplet of mandibular gland secretion at the center of the concentric circles shown in Fig. 4.16. In still air each component diffuses away from the site of deposition. Because they differ in volatility, some fill spece more quickly than others. Sfter a short time, hexenal defines the largest active space, indicated by the outermost circle. When ants enter this space, the alerting behavior is characteristic. As they move into the next zone (hexanol), the major workers are attracted toward the site of its deposition. In the innermost zone 2-butyl-2-octenal is the biting marker, and 3-undecanone contributes to biting and is also a short- range orientation signal. When the secretion has been allowed to evaporate for some minutes, no effects of hexanal or hexanol are detectable, but as soon as the workers come close to the deposition

Hexanal

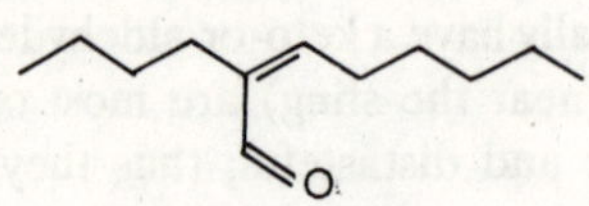

2-butyle-2-octenal

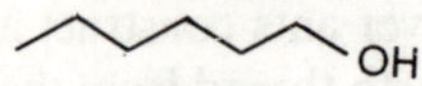

Hexanol

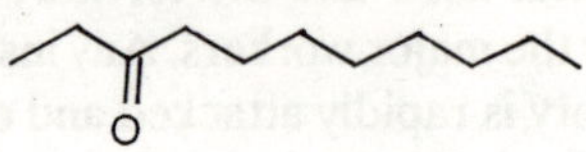

3-undecanone

site, biting behavior is released. Consequently a chain of behaviors is sequentially triggered by the discrete chemical components of the alarm pheromone that are localized on the intruding object.

As opposed to the mayor worker ant the minor workers of the weaver ant emit a considerably different set of components from their mandibular glands when they are disturbed. If other minors are exposed to that secretion, they do not defend the nest but instead flee from the site of disturbance. In contrast, major workers are attracted and arrested by high concentration of at least one component in the minor's perfume. Thus major workers come to the defense of the distressed minor worker and of the nest.

How much different is one alarm signal to the other? Are they specific for each ant species or for each colony? Apparently they are much less so than for sex pheromones. Often the same components are shared among many species within a genus and among genera within a subfamily. How- ever, there are subtle and biologically important differences between similar species, as shown by studies in the ant genus *Myrmica*. Four species of this genus that coexist in similar habitats in England and Belgium have been found to have similar blends of several components in their alarm pheromones, but their relative proportions differ. In effect, each species has a different form of communication.

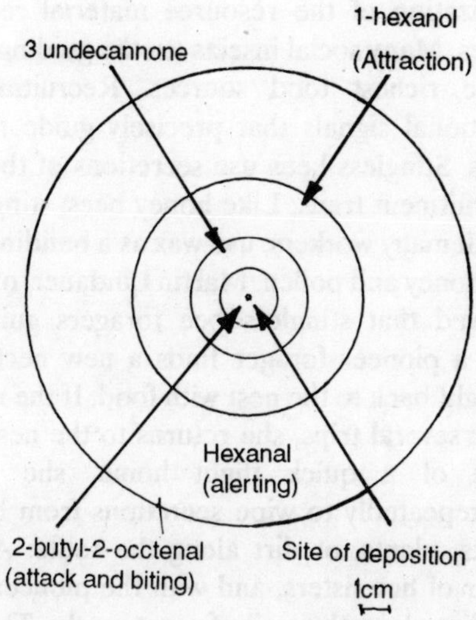

Fig. 4.16. The mandibular gland secretions of major workers of weaver ants include four components with specific behavioral effects. These four differ in their volatility, and at 20 seconds after it as been deposited, the mixture creates a set of concentric active spaces.

Trail and Territorial Markings.

Both food and habitat are restricted. Territorial and trail pheromones enable efficient resource exploitation. Many foods, like blue-berries or caterpillars, act as the container of insects eggs. Female insects that place their eggs in such organic matter and mark them with a pheromone that is like a sign that says "Occupied." Some parasitic wasps include such a territorial phero- mone with each egg injected into an insect host. Female apple maggots mark each fruit after ovipositing in it. After she has laid an egg, the female drags her ovipositor over the surface of the fruit and thus deposits pheromone. The size of the mark depends on the future needs of one larva from hatching to maturity. A small fruit, like a cherry, is completely coated after one oviposition and thus will be avoided by subsequent females. A large fruit, like an apple, is marked on only a portion of its surface. It will go on collecting several eggs until its surface coat of pheromone reveals that it is loaded to capacity with larvae.

Effective utiization of the resource material requires choices among alternatives. Many social insects usethe guiding trails to direct nestmates to the richest food sources. Recruitment trails are continuous directional signals that precisely guide the foragers to abundant goodies. Stingless bees use secretions of their mandibular glands to lay recruitment trails. Like honey bees, stingless bees have large colonies with many workers, use wax as a building material, and collect and store honey and pollen. Martin Lindauer, of the University of Munich, showed that stingless bee foragers guide each other efficiently When a pioneer forager finds a new nectar source, she makes a direct flight back to the nest with food. If the nectar source is still copious after several trips, she returns to the nest in a different manner. Instead of a quick flight home, she flies in small stages–stopping repeatedly to wipe secretions from her mandibular glands onto rocks, plants, or dirt along the route. At the nest she arouses a number of her sisters, and with the pioneer as leader and pilot, the group flies along the trail of scent marks. The foragers stop along the way to check the marks, and finally arrive at the source of rich nectar.

The tacticswhich guide these stingless bees require both the pilot bee and an odor trail. In many ant and termite species an odor trail alone is sufficient to guide a follower to a food source.

In thge case of ants and termites the trails are laid by dragging their abdomens along the substrate and depositing streaks of chemicals. The active space of the trail is a narrow corridor and extends upward in a half-cylinder. Followers run along within this active space, weaving from side to side and turning back to the center as their waving antennae break out of the active space into an unscented zone. The followers do not have to touch the odor marks but have only to stay within the active space. They can even follow a trail drawn overhead on the ceiling if ceiling and floor are not too much apart.

The function of these trails differ accoding to the biology of each species. In harvester ants, Bert Holldobler, of Harvard University, has shown that some are fixed trunk trails while others are more temporary recruitment trails. Trunk trails, laid by secretions of Dufour's gland, are like spokes radiating out from the nest entrance into foraging territory. Trunk trails are fairly permanent and are used mostly by ants that are returning to the nest. Recruitment trails, laid by the poison gland, branch off from trunk trails. They are laid by a successful forager as she returns from a food source. These trails entice other foragers to follow the chemical signals and thus to help retrieve the food. As food-laden foragers come back to the nest, they lay recruitment trails that reinforce the previous signals. Foragers that do not find food and return hungry to the nest lay no recruitment trails. The recruitment trail has a short half-life, it gradually fades away along with the finishing of the source of the food.

It is apparently easy to say that a given species of ant or termite follows a chemical trail, but the molecular components of the trail signal have proved difficult to isolate in most cases. To demonstrate the presence of a chemical trail, a researcher need only paint s line of a chemical sample along a substrate, such as a piece of filter paper, and then watch to see whether foraging ants or termites follow the line when they chance to cross it. Extracts of whole insects or their glands or chemical composition can be esily analysid.

Not more than a dozen trail pheromone components have been identified, and most of these are not very volatile. Neocambrene A, a complex terpene, is the trail substance of nasute termites, and a pyrazine is the trail substance in several species of myrmicine ants. At least in these ants, the true trail pheromone is not species specific. This pyrazine is found in all seven species of British *Myrmica*, and it is found also in a related genus of tropical leaf-cutter ants. The power of modern microchemistry is such that only 50 *Myrmica* workers provided enough pyrazine for the adequate examination of the chemical composition of the trail.

The poison gland produce trail pheromone of *Myrmica*. When Marie-Claire Cammaerts, of the University of Brussels, Belgium, drew lines with poison gland extracts, workers that blundered across the lines turned to follow them, but foragers were not attracted to the trail. Further experiments revealed that the volatile components of Dufour's gland are attractants and excitants. Both glands have a common opening at the tip of the abdomen and both are responsible in forming the trail.

In the natureal habiat trails are not public thoroughfares that are shared commonly by all species. Each species, and usually each nest, follows its own trails. An artificial trail, laid with a combination of the pyrazine and volatile components of the Dufour's secretion of one

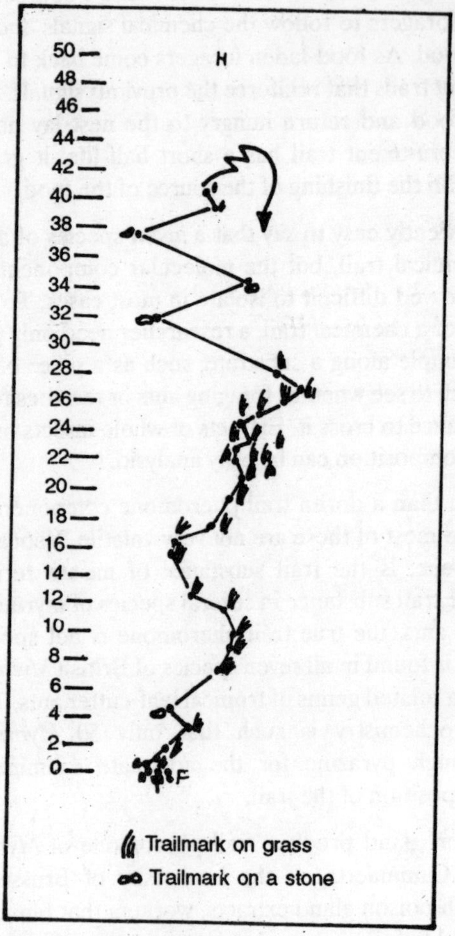

Fig. 4.17. The record of a trail-marking flight of *Trigona postica*, a stingless bee. Each point indicates one of the short landings on her return flight, during which she deposited an odor trail mark. *H*, hive; *F*, feeding table.

species of *Myrmica*, is followed quite well by foragers of the samw species and equally ewll by foragers of other species. There is no obvious species identification. However, when Cammaerts added the remainder of the Dufour's secretion, which is an oily mixture of low volatility, she produced trails that were species specific. The oils differ among the species and allow each species to identify its own unique trail.

The secreted oils from Dufour's glands are a ***territorial pheromone*** for *Myrmica*. Pioneer foragers, seeking new food sources far from the nest, move forward slowly and tentatively. As they walk ahead onto new ground, each forager repeatedly marks the ground with her Dufour's gland. The oils persist for long periods, and subsequent foragers will run quickly across the same ground that was so slowly crossed the first time. The oily marksdiffferentiate home territory from foreign territory.

The natural foraging trails of *Myrmica* have three kinds of components, each of which serves a distinct purpose. Secretions of the poison gland define the trail itself. Volatile substances from Dufour's gland attract and increase running speed, but these signals are short-lived. Finally, oily substances from Dufour's gland survive and mark the home territory.

Pheromones and Individual Differences.

A few members of species do not emit the same complement of pheromones. Differences between life stages (larva versus adult) and between sexes are obvious. Also, genetic strain, physiological state, and nest and caste identification in social insects are identifiedk by pheromones.

For example, for successful mating of the fruitfly, *Drosophila melanogaster*, the female must emit pheromones that stimulate the male to court, and the male must emit pheromones that cause the female to accept. Both sexes prefer to mate with "strangers" that are genetically quite different, and they recognize the differences between related and unrelated partners by their differing pheromones. Males and females from the same inbred line do not spontaneously court one another unless they are fooled by being brought together in an environment filled with alien pheromones.

Most of the halictine bees are colonial and nest together in burrows. At the entrance to each colony, a guard bee inspects newcomers and rejects those that are not residents. Charles Michener and his associates at the University of Kansas wondered how the guards of one species. *Dialictus zephyrus*, distinguished nestmates from aliens. Bees were brought to the nest opening in a piece of plastic tubing and poked out into the entrance with a pipe cleaner. Of the 228 nonresident bees introduced into other nests, 50% were attacked by guards before contact and 92% on contact. Fifty residents were introduced in a similar manner, and all were accepted. Then, Michener killed 20 residents and 32 nonresidents by freezing and presented them, after thawing, one by one to the guard bees. Merely 10% of the frozen residents were rejected, while 97% of the nonresidents were attacked. Nonresidents were also rejected when the room was illuminated with a red light to which bees are blind. Thus,it is confirmed by these experiments and others established that touch, vision, and hearing are not used for rejection. Odor must account for recognition of nestmates. Does the guard remember the scent of a nestmate, or does she merely compare the odor of the entering bee with that of the nest? When guards are placed on duty in a foreign nest, they accept only their own nestmates and reject the owners of that nest. It is the odor of the bee, not that of the nest, that these guards use to detect the identilty of the intruder.

The scent if the colony has a strong hgenetic component. Nestmates are sisters and are genetically similar. Guard bees of *Dialictus zephyrus* can recognize individual odors that are variants on the pheromone blends produced by their nestmates, and they will thus admit genetically related bees that were reared in other nests. In honey bees. Even in the case of honey bees, colony odor has a genetic basis. When a honey bee queen is suddenly presented to workers of another colony, she is usually attacked. But she is much more likely to be accepted by these workers if she is closely related to their own queen. Michael Breed, of the University of Colorado, Boulder, has shown that inbred sister queens are accepted in 35% of the exchanges, while the queen bees which are unrelated are definitely not accepted.

Conditioning plays a part in the recognition of individuals or classes of individuals. In laboratory colonies, an halictine guard bee, placed in a "foreign" nest, can habituate to the odors of that nest and then will accept a bee with that odor, but such a situation is probably rare in nature. The response of male halictines to females is more meaningful. Upon first scent, a male responds avidly to the novel odor of a new female. But if she is unwilling to be courted, he becomes unresponsive to her now familiar odor. When exposed to a second new female with yet a different odor, he immediately attempts to court again. It is clear that the malw bee learns and accepts quickly the response of lthe female bee.

Probably the glands which emit a distinct body odour are many. The surface layers of insect cuticles contain odorous molecules that originate in the underlying epidermal cells. Other contaminants are left on the surfaces around the exocrine glands and dissolve into the coat of epicuticular and surface lipids. Each exocrine gland has many components, and their proportions vary slightly between individuals. The rich mixture of secretions from many glands, the scents from the environment, food odors, and the intrinsic odors of the epicuticle combine to produce pheromonal polymorphism enabel each insect with ujnique identity.

Primer Pheromones.

Primers bring changes in the physiological state of the recipient. It is usually assumed that primer effects are mediated by the endocrine system. Thus pheromones influence endocrine glands, and hormones directly regulate the physiological state. Primers are found in subsocial and social situations where the persistence of the group allows stimuli produced by one insect to influence another.

Most primers influence reproductive capacity. A pheromone from male desert locusts increases the rate of development in immatures of both sexes. The scents of male and female mealworm beetles accelerate egg development and other aspects of reproductive maturation in young female beetles. Honey bee queen substances stop the growth of ovaries in the workers.

It is quiet interesting to note that the primers that switch development from one end point to an alternative. Desert locusts have such a primer, locustol, which triggers development from the solitary form to the gregarious form. The gregarious form is responsible for the devastating locust swarms that have ravaged crops in Africa since biblical days. As more and more locusts feed together, they begin to produce locustol in their feces. Locustol causes a gradual transformation of the light-colored solitary types into much darker gregarious forms that gather together and migrate, devuring all that comes in thir way.

Generally the colonies of termite have only two functionally reproductive individuals at any time. If an accident or an entomologist removes the male or female reproductive, one of the nymphs of the same sex develops into a "supplementary reproductive". The presence of a functional reproductive inhibits the maturation of others of the same sex. It is presumeably each reproductive insect produces an inhibitory pheromone, which is passed around by mutual grooming to workers and among workers. In at least one species of termite, it is possible to inhibit sexual maturation of nymphs by introducing extracts of the functional reproductives. Soldier production is probably limited by an analogous soldier-specific inhibitory primer, but there is no experimental evidence to conform this

Defensive Allomones

Most insects prevent the possibility of being eaten because they are distasteful. Foul-smelling and eviltasting chemicals in both animals and plants confer proection from attackers.

Probably the easiest chemical defensive behavior for an insect is regurgitation-vomiting gut contents at an attacker, as do many grasshoppers when they "spit tobacco." The regurgitate may be enriched in unpleasant chemicals derived from the host plant. For instance, sawfly larvae that feed on pine trees sequester terpenoid resins from the needles in two pouches of the foregut. The purches, like the rest of the foregut, are lined with a cuticle that prevents diffusion of the resins into the blood of the insect. When attacked by an ant or a spider, the sawfly rears up its front, regurgitates a droplet of resinous fluid, and wipes the regurgitate onto the attacker. The regurgitate is rich in the resins of the pine tree, which are repulsive and distasteful to birds and so arthropod predators such as ants and spiders.

Blood-Borne Allomones.

Most of the unpalatable insects are aposematic (warnings colored), such that predators can readily associate the appearance of the animal with its unpalatability. For example, monarch butterflies are strongly aposematic with their striking orange and black wing patterns,. When blue jays try to eat monarch butterflies, they vomit up the partially eaten carcass (Fig.). Linclon and Jane Van Zandt Brower, then at Amherst College, Massachusetts, offered moncarch and other butterflies to captive jays. The birds quickly learned to associate the monarch color pattern with vomiting, and thereafter the b;utterflies were reduced. Moncarch color pattern with vomiting, and thereafter the b;utterflies were refused. Monarch butterflies are unpalatable because their blood and tissues contain cardiac glycosides. These glycosides were sequestered by the caterpillars from milkweed in their diet.

The insects which carry the defensive allomone in blood, it may be liberat a special localized bleeding, or autohemorrhage. Blood bubbles out from weakened sites at leg joints, antennal joints, or between sclerites when the insect is attacked. Rough handling of a blister beetle will cause it to release clear droplets of blood from its leg joints, and lady beetles exude yellow drops of blood from similar sites (). The blood of the blister beetle contains cantharidin (spanish fly), while that of the lady beetle carries the alkaloid precoccinelline. In contrast to the cardiac glycosides used by monarch butterflies, both cantharidin and precoccinelline are manufactured by the insect itself rat her than extracted from a food plant.

The substance such as, cantharidin and precoccinelline are distasteful to many vertebrate predators, but not to all. Thomas Eisner reports that toads will repeatedly consume blister beetles. Small predators are further deterred by the mechanical properties of the droplet. Ants become entangled in the sticky blood. As it coagulates, one leg is glued to another, antennae become stuck to mouthparts and ants adhere to each other. Ants retreat off the attack and go through a frenzy of cleaning activity.

Allomones Produced in Exocrine Glands.

Defensive allomones are frequently found in cuticular reservoirs (as were those of the sawfly). Most often the reservoirs are storage compartments for secretions made within exocrine glands. The reservoir may be eoerted and wiped on the site of attack, as seen in the larvae of swallowtail butterflies . The defensive fluid of these caterpillars con- tains isobutyric and 2-methyl butyric acid. Alternatively, the secretion may ooze from the gland opening, much as

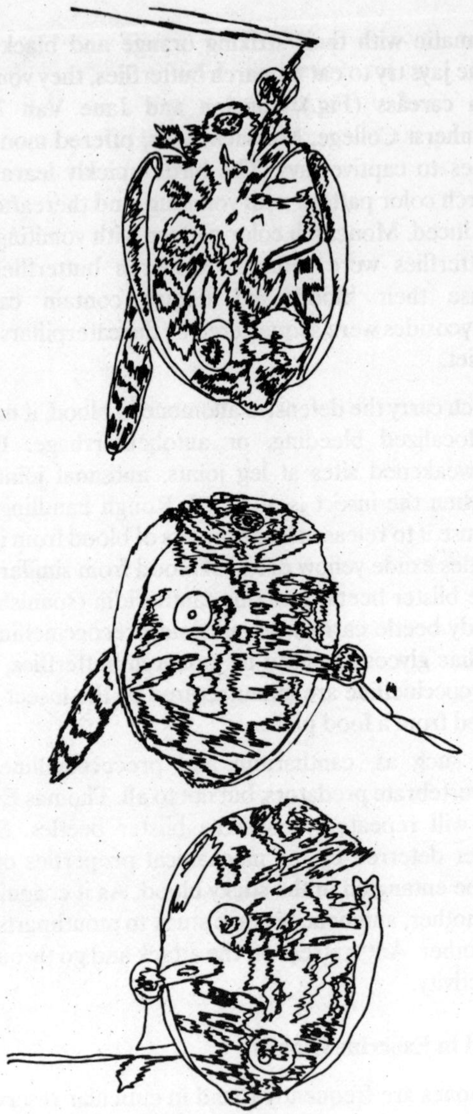

(a)

Fig. 4.18. When roughly handled, lady beetles (a) and blister beetles (b) release droplets of blood from their leg joints. The droplets contain distasteful chemicals. Note that when a leg is probed with forceps, lady beetles release droplets from that leg.

blood oozes from the knee joints of blister beetles. Finally, the secretion may be ejected in a jet of acrid mist. The spraying glands are dramatic in their impact and often extremely accurate. The defensive fluid of a Florida walking stick is aimed directly at the site of attack and effectively repels both vertebrate and invertebrate predators.

Cuticular reservoirs increases the chances of potential defensive allomones, for within a sac of impervious cuticle, small reactive molecules can be safely stored. In contrast, these toxins cannot be stored in the blood.

The chemical substances wihich protect the insects operate in a very different context from that of chemical signals such as

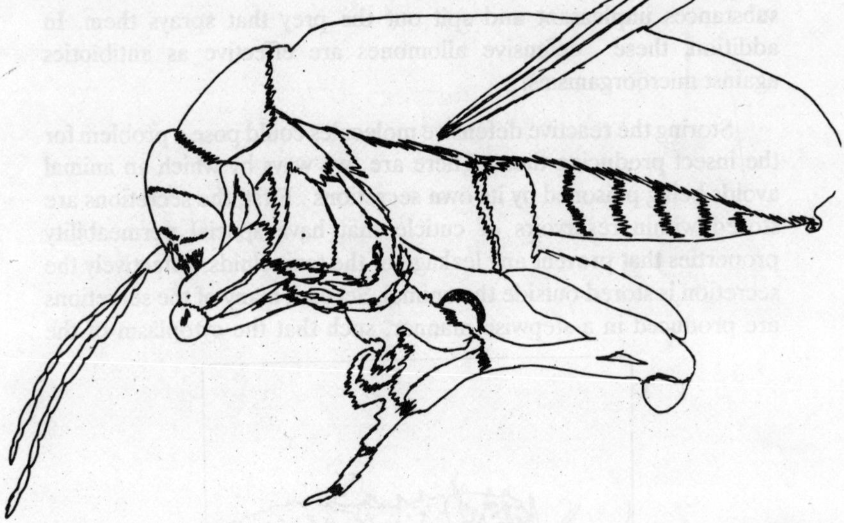

pheromones. The potential prey is at an advantage when the signal is so blatant that the predator cannot ignore it. The predator is at an advantage when it can ignore the blatant insult or when it is insensitive to the defensive substance. Any insect has a host of potential predators, and one might expect natural selection to favor defensive allomones that are nonspecific. That, is are the repellant to many species. The oefensive secretions of many insects contain small reactive molecules that are broadly toxic to most living tissue. Small acids, like formic or acetic, readily coagulate protein and irritate excitable tissues. The chemical sbustances. such as formic and acetic acids are used in the defenses mechanism of ants, beetles, and caterpillars. Reactive adlehydes, which bind readily to proteins, are painful tear gases to humans and animals. Molecules such as these are found in the defensive secretion of walking sticks (*Anisomorpha*), cockroaches, beetles, and true bugs. Benzoquinos, which also bind proteins and "tan" them, are common in the defensive secretions of beetles and millipedes. Many predators findsuch small reactive substances unpleasant and spit out the prey that sprays them. In addition, these defensive allomones are effective as antibiotics against microorganisms.

Storing the reactive defensive molecules could pose a problem for the insect producing them. There are two ways by which an animal avoids being poisoned by its own secretions . First, the secretions are stored within reservoirs of cuticle that have special permeability properties that prevent any leakage of the toxic fluids. Effectively the secretion is stored outside the animal. Second, many of the secretions are produced in a stepwise manner, such that the cytoplasm of the

Fig. 4.19. Caterpillar of a swallowtail butterfly extruding its osmeteria in response to a disturbance.

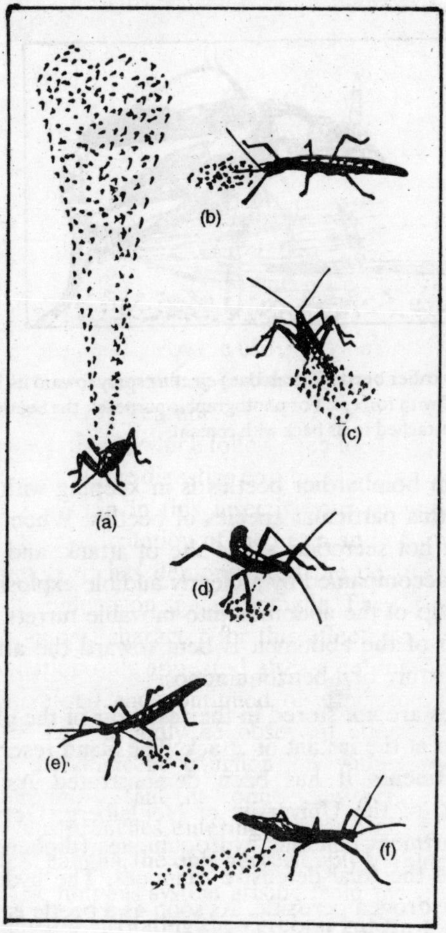

Fig. 4.20. *Anisomorpha*, a walking stick, spraying in response to stimuli applied in various ways. (a-c) Bilateral discharges elicited by tapping the dorsal thorax (a) touching both antennae with a heated probe (b), or pinching the rear of the abdomen (c). (d-f) Unilateral discharges elicited by pinching the right foreleg (d), left middle leg (e), and right hind leg (f).

secretory cells contains only harmless precursors rather than the toxic end product. The progression of the production machinery is best seen in the adaptation for quinone production by bombardier beetles and darkling beetles.

Fig. 4.21. Bombardier beetle (Carabidae) ejecting spray toward its left front leg, which is being pinched with forceps. For photographic purposes, the beetle has been fastened to a wire hook attached to its back with cement.

The term bombardier beetles is in keeping with the defensive behavior of this particular species of beetles. When disturbed, they emit a jet of hot secretion at the site of attack, and the ejection of secretion is accompanied by a clearly audible explosion. The glands open at the tip of the abdomen into movable turrets that are aimed when the tip of the abdomen is bent toward the attack The spray contains a mixture of *p*-benzoquinones.

Quinones are *not* stored in the reservoir of the gland but rather are produced at the instant of attack. The gland reservoir consists of two compartments It has been demonstrated As shown by H. Schildknecht, of the University of Heidelberg, Germany, that the inner compartment contains hydroquinones (diphenols), which are precursors to the final defensive quinones. The hydroquinones are mixed with hydrogen peroxide. As soon as a beetle is attacked, some of the content of the inner compartment is passed into the outer compartment. The author compartment is a reaction chamber. The small outer compartment contains a mixture of enzymes (catalases and peroxidases) that oxidizes and decomposes the hydrogen peroxides to give water and oxygen and also oxidizes the hydroquinones to yield benzoquinone. The oxidation reactions liberate heat and gaseous oxygen. Thomas Eisner has shown that the temperature of the mixture reaches 100°C. The gaseous oxygen provides the propellent that accounts for the audible "pop" when the glands discharge. Eisner has also used high-speed photography to analyze encounters between attacking ants and bombardier bettles.

Ths entire sequence, from ant bite to beetle discharge to ant release, takes place within a second.

Another species known as Darkling beetles, or flour beetles and mealworm beetles, also use quinones in their defensive secretions. Unlike the bombardier beetles, these darkling beetles store the actual quinones, not a precursor, in the reservoir of the glands. But like the bombardier beetles, the darkling beetles protect their secretory cells from quinoid products by a strategic sequence of localized reaction steps. The cells which secrete these defensive glands pass their products to the reservoir via fine cuticular ductules. In these defensive glands there are two distinct populations of secretory cells. The secretory cells that are of particular interest to quinone production are types 2a and 2b, which are arranged in pairs. Each pair is strung in series on a cuticular ductule that leads to the reservoir.Each cell of the pair surrounds a central cavity within which lies the efferent ductule. The secretory product of the cell passes into that extracellular cavity and thence through the wall of the efferent ductule and finally to the reservoir.

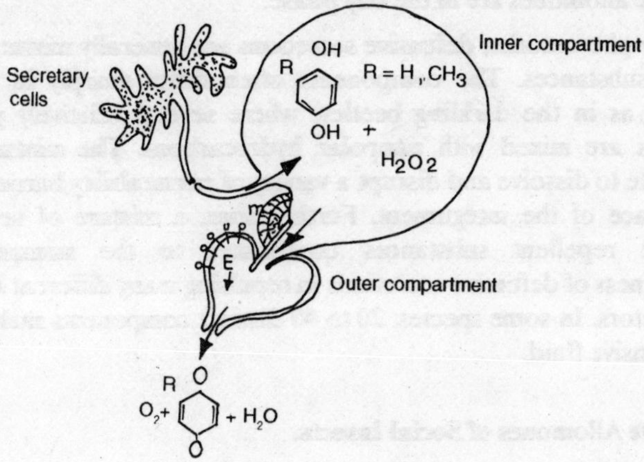

Fig. 4.22. Diagram of one of the paired defensive glands of the bombardier beetle. The secretory cells produce a mixture of hydroquinones and hydrogen peroxide, which are stored in the reservoir. When a muscle (m) opens the valve, some of the contents of the inner compartment flow into the outer one, where catalases and peroxidases catalyze an explosive exergonic reaction.

The structure of cell cytoplasm, cavity, and ductule can be viewed as reaction compartments, arranged in series. The two final steps in quinone production are hydrolysis and oxidation. The reaction steps for quinone production apparently occur in the sequential compartments. In the cytoplasm of the distal cell, phenyl glucoside is stored. The glucoside is hydrolyzed as it is secreted into the cavity of the distal cell, and the resulting diphenol can then pass along through the wall of the ductule. Within the ductule, the diphenol is oxidized to produce the final quinone. The oxidative enzymes (phenolases and peroxidases) are apparently built into the wall of the cuticular ductule. Thus the secretory cells avoid being poisoned by quinones because they are never actually exposed directly to poison.

Another mechanism, to which protects insects from the poision they develop themselevs gradually to secrete appropriate enzymatic catalysts and precursors of the allomones into thereseervoir, and then to allow the final steps to take place in the reservoir. Reservoir fluids often consist of a watery phase and an oily phase. Murray Blum and his colleagues report that in the two-phase secretion of a true bug (*Leptoglossus*), the enzymes are in the aqueous phase and the reactive defensive allomones are in the oily phase.

Like pheromones, defensive secretions are generally mixtures of several substances. The components often differ sharply in their polarity, as in the darkling beetles, where several relatively polar quinones are mixed with nonpolar hydrocarbons. The mixture is tadequate to dissolve and disrupt a variety of permeability barriers at the surface of the integument. Furthermore, a mixture of several different repellent substances contributes to the nonspecific effectiveness of defensive secretions in repulsing many different kinds of predators. In some species, 20 to 40 distinct components make up the defensive fluid.

Defensive Allomones of Social Insects.

As soon as the nest of a social insect is invaded, members of the colony pool their resources for collective defense. In some ants and many termites, a special caste, the soldiers, is dedicated to nest defense. The secretions they use are doubtly effective repellent to the attackers and alarm signals for nestmates. As shown by endure Quennedey at the University of Dijon in France, the major source in higher termites is the frontal gland of the soldiers In *Cubitermes*, the strong mandibles pinch the attacker (usually an ant), and the frontal gland secretions pour out on the sites of contact. The secretions are somewhat repellent, but in this genus the mandibular pinch is the major defensive act. In *Rhinotermes* there are two caterogries of soldier: a major caste with enormous mandibles and a minor caste with tiny mandibles. The ninor caste has a huge frontal gland reservoir

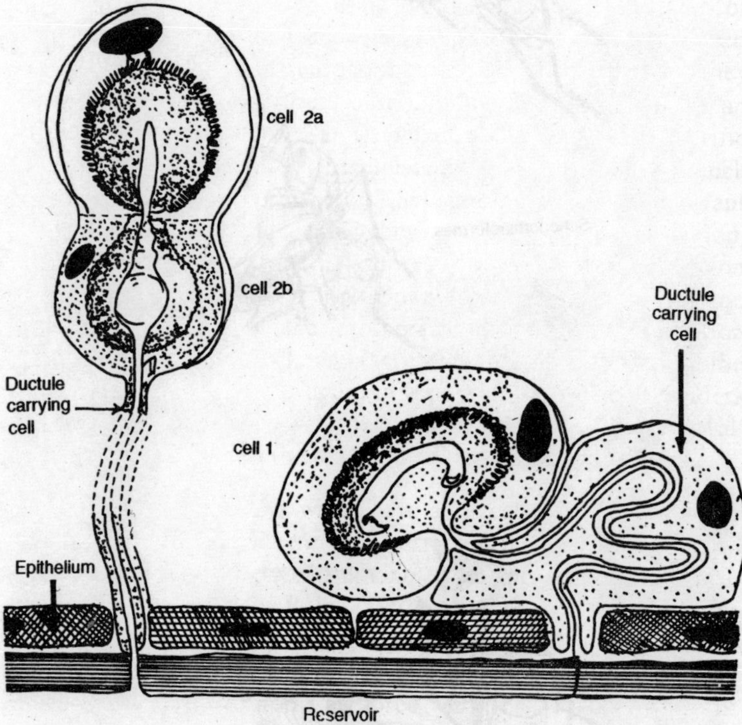

Fig. 4.23. Secretory cells in the defensive glands of darkling beetles (Tenebrionidae). In the type 1 secretory units, a single secretory cell is drained by a cuticular ductule. In the type 2 secretory units, two secretory cells are arranged in series along cuticular ductule. In an entire gland there are hundreds of each type of secretory unit, all of which pass their products to the reservoir.

that extends back from the head all the way to the abdomen. It secretes a mixture of ketones toxic to many insect species. In the third style of defense, found in nasute termites, the chemical weapon is paramount. The soldiers are remarkable in that their mandibles are

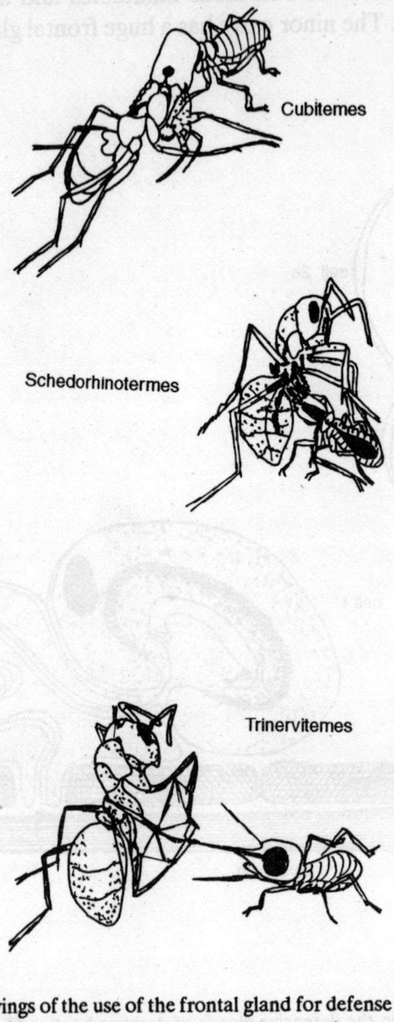

Fig. 4.24. Drawings of the use of the frontal gland for defense in three species of higher termites. In *Cubitermes*, the secretions supplement the strong mandibular pinch. In *Schedorhinotermes*, the minor soldiers have a very large defensive gland extending back into the abdomen. The defensive secretions of nasute soldiers, like *Trinervitermes*, entangle the attacker

reduced in size and their heads are huge, elongated snouts-squirting devices for the frontal gland secretion. When the nest wall is broken, soldiers advance outward in parallel or in staggered line, precisely spaced by just the distance of antennal contact. A soldier sprays the frontal gland contents when it is bumped. The secretion contains volatile terpenes that are alarm releasers for termites and irritants to the attacker, and other unusual terpenes that are sticky and entangle ants, which are major threat to the lives of termites.

Allomones Promoting Associations

Defensive allomones terminate interspecific associations between potential prey and a diverse array of predator species. In contrast, there are a host of less-studied allomones that promote intimate associations between two particular species. Many of these associations are mutually aggreable with both partners deriving substantial benefits, while others are parasitic, when only one of the partner gerts the benefits.

A strange phenomenon is seen in the behaviour of specialized insect species or "guests" in the nest of social insects. The alien insect enters an ant nest or termite nest without being attacked and is sometimes even assisted by a host worker. In some cases the volatile body lipids of guest and host are very similar in composition so that the intruder appears to mimic the body odor of the hosts. In other cases the guests placate their hosts with allomones that E. O. Wilson has termed ***appeasement substances.*** Jacques Pasteels, of the University of Brussels, Belgium, has shown that the appeasement substances of rove beetles are produced in a battery of special exocrine glands that are found only in species that are social parasites. Bert Holldobler has observed the process by which a rove beetle guest grains entry to an ant nest. When a forager of *Myrmica* approaches a beetle, the beetle promptly bends its abdomen forward and presents a secretion to the ant. The ant feeds on the appeasement secretion. Next, the ant licks an "adoption gland" on the lateral margins of the beetle's abdomen, and finally it welcomes the beetle into its nest.

when inside the nest some guest species behave generously enough, merely soliciting regurgitated food from the workers and exploiting the shelter and stable habitat provided by the hosts. But some guests return the hospitality with a bad turn. Kenneth Hagen, of the University of California, Berkeley, has reported that some

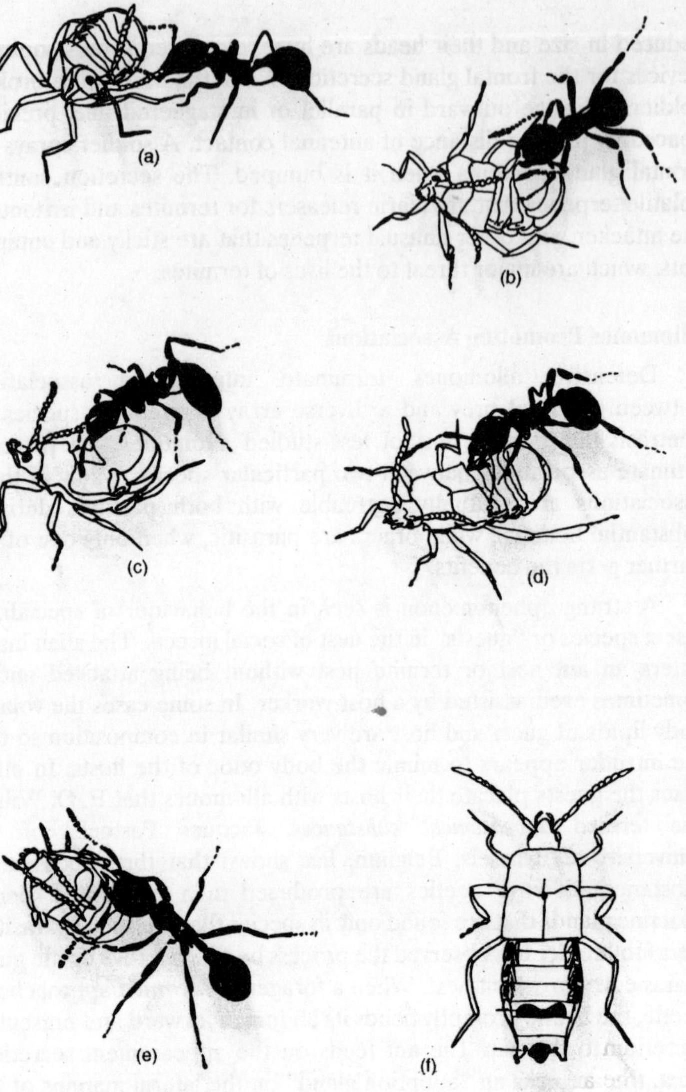

Fig. 4.25. A rove beetle (Staphylinidae) (white) being taken into the nest by a worker ant (black). The beetle presents its appeasement gland to a *Myrmica* worker that has just approached it (a). After licking the gland opening (b), the worker moves around to lick the adoption glands (c-d), after which it carries the beetle into the nest (e). In (f), ag = adoption glands; dg -{defensive glands; apg = appeasement glands.

lacewing larvae eat their termite hosts. The lacewing larvae move about freely in the termite galleries. At certain developmental stages the acewings approach termite workers and wave the tips of their abdomens in front of the termites. The termites are paralyzed by a gaseous allomone, and the larvae then devurs the paralysed workers.

Allomones also promote symbiosis between insects and micro-organisms. Leaf-cutter ants, certain termites, and bark beetles transport and consume symbiotic fungi. The associations are quite specific–each insect species cultures only one or two fungal species. As the fungus grows in ant nests, the cultures remain clear of any contaminating species. Leaf-cutter ants weed their crops by mechanical and chemical means. Alien spores are picked out of the fungus garden, and allomones, produced by the metapleural gland of the thorax, regulate the species composition of the growing fungi. H. Schildknecht and V. Maschwitz reported that the metapleural secretion has three components: phenylacetic acid, β-indolyacetic acid, and myrmicacin (β-hydroxydecanoic acid). Phenylacetic acid is a general-purpose antibiotic that prevents the growth of bacteria and of some weed fungi. Indolyacetic acid, a plant hormone, is thought to promote growth of the symbiotic crop. Myrmicacin inhibits the generation of fungal spores. The species of fungus normally cultured in colonies of these ants is unaffected by phenylacetic acid or myrmicacin.

Both, Bark beetles and ambrosia beetles transport symbiotic fungi within cuticular pits or in large cuticular pockets. The pits and pockets are associated with secretory cells (Fig. 4.2 to 4.4), and cycles of secretary activity correlate with the peak of fungal proliferation in the pocket. In the case of the southern pine beetles, the secretions apparently nourish the symbiotic fungi and restrict their growth to yeastlike budding rather than production of long hyphal strands. The yeastlike progeny drop out of the transport pocket as the beetle asses through the tunnels and galleries and start new growths of symbiotic fungi that will nourish beetle larvae. In addition, the microflora in the pocket is restricted to two species. Presumably the secretions favor the growth of these two symbionts while excluding all the weed fungi that contaminate the integument of the beetle. Unfortunately we unfamiliar of the chemical nature of allomones that play these role.

The mircorsymbiont of the association may be an aid to digestion. As shown by the late L. R. Cleveland, of Harvard University, intestinal protozoans (flagellates) help the primitive cockroach *Cryptocercus* to digest its meal of cellulose. The protozoans live in the anaerobic hindgut of the cockroach. In the presence of high concentrations of oxygen, the protozoans die. To grow, the cockroach must molt its entire cuticle, including that of the hindgut. If abruptly exposed to the atmosphere together with the shed skin, the protozoans would not survive, and the cockroach would lose its vital digestive parts

The protozoans save themselves from a molt because they "anticipate" it and enclose themselves in a cyst that excludes poisonous oxygen. The allomone that triggers encystment is the insect hormone that induces apolysis: the familiar steroid ecdysone. The newly molted cockroach consumes its old cuticle, and once within the anaerobic gut environ- ment, the protozoans become activates again.

An Overview of Chemical Signals

There has been chemical signals between cells since the moment first cells appeared on the earth, and intercellular chemical communication is a distinguishing feature of all major phyla. Hormones, pheromones, and allomones share some common originsin the earlier phase of the evolution of lifeform on the earth. Exocrine secretions for diverse purposes often have metabolic byproducts, and it is not surprising that these products, which are produced only in certain physiological states, might communicate that state to animals of the same species. For pheromones are those allomones that promote associations, selective pressures act on both the producer and the target organism to evolve an unambiguous signal that is detected at low levels. For defensive allomones, selection operates on the producer to make the signal blatant and insulting, while the potential target organisms are at an advantage if they can ignore the insult. Thus many defensive allomones are broad-spectrum reactive protein poisons that are vitally important in the oves of insects.

Same kind carbon chains form the skeletons of the molecules that function as allomones, pheromones, or hormones. The same molecule may play more than one role, as does ecdysone in *Cryptocercus*. Theoretically structural diversity of chemical signals should be almost unlimited, but in practice only a few families of carbon skeletons predominate among pheromones, hormones, and allomones. These include aliphatic straight chains, terpenes and the related steroids, and small ring compounds.There are also a significant number of hormones are peptides, and a significant number of defensive, allomones are complex and unusual poisons. The explanation of the widespread molecular conservatism is probably rooted deep in the metabolic pathways common to living systems. All cells have the enzymes for manufacturing certain skeletons. To develop a signal molecule, it is simpler to amend a preexisting biosynthetic pathway and a preexisting receptor surface than to evolve totally new ones.

All the categoris of molecular signals have common characteristics. Mostly phoromones and many allomones are volatile. Many defensive allomones are simple poisons with quite reactive functional groups, while others are complex and unusual toxins (such as precoccinelline) that irritate sensory receptors and are difficult to degrade metabolically. Most pheromone components are scarcely reactive, but their skeletons tend to be rigid and to have asymmetric carbons that determine precisely the shape of the molecule. Both defensive allomones and pheromones are medleys, but for very different reasons. Defensive allomones are medley, but for very differemt reasons. Defensive allomones are medleys so that the producing animal covers as many bets as possible when it plays its defensive hand. Pheromone medleys allow the construction off fever number of indifiniteness signals.

Summary

Chemical communication between insects requires an exocrine gland, its chemical product, and an individual that receives the signal. Pheromones are chemical signals that act between members of the same species. Allomones act on members of another species.

Most of the familiar pheromones are releasers that elicit a prompt behavioral response. The chemical signal spreads out from the source to create an active space a zone within which other organisms respond. The size and the life span of the active space vary with different types of signals.

5

Incompletely Known Substances with Allegedly Hormonal Characteristics

The dorsal (thoracic) glands of Lepidopteran larvae

In a number of research papers, Hintze (1968, 1969) and Buckmann (1967) described a pair of ductless glands in the dorsal region of the pro- thorax and mesothorax which the latter author distinguished with the mandibular glands of other insects Each followed the main ipsilateral dorsal tracheal trunk. Their anterior parts were formed of two branches – one terminating on the head apodeme of the mandibular musculature and the other coming into contact with the corres- ponding corpus allatum. The unbranched posterior parts entered the fat body, where they were attached dorsolaterally to the epidermis. The characteristic endocrine character of the glands was demonstrated in larvae of *Cerura vinula* and *Notodonta anceps*.

In these type of species, the anterior, thinner portion of each gland was formed of closely packed cells, giving a smooth surface. Evidently thicker posterior portion have two rows of large, round polyploid cells resembling the prothoracic glands of Lepido- ptera in their large nuclei and secretory cycles. Hintze (1969) stated that the anterior portion produced lipoproteins and the posterior portion lipids. In the other Lepidoptera studied (*Sphinx ligustri, Smerinthus ocellatus, Lymantria monacha* and *Malacosoma neustria*), however, the mandibular glands had a well-developed duct lined with a chitinous intima.

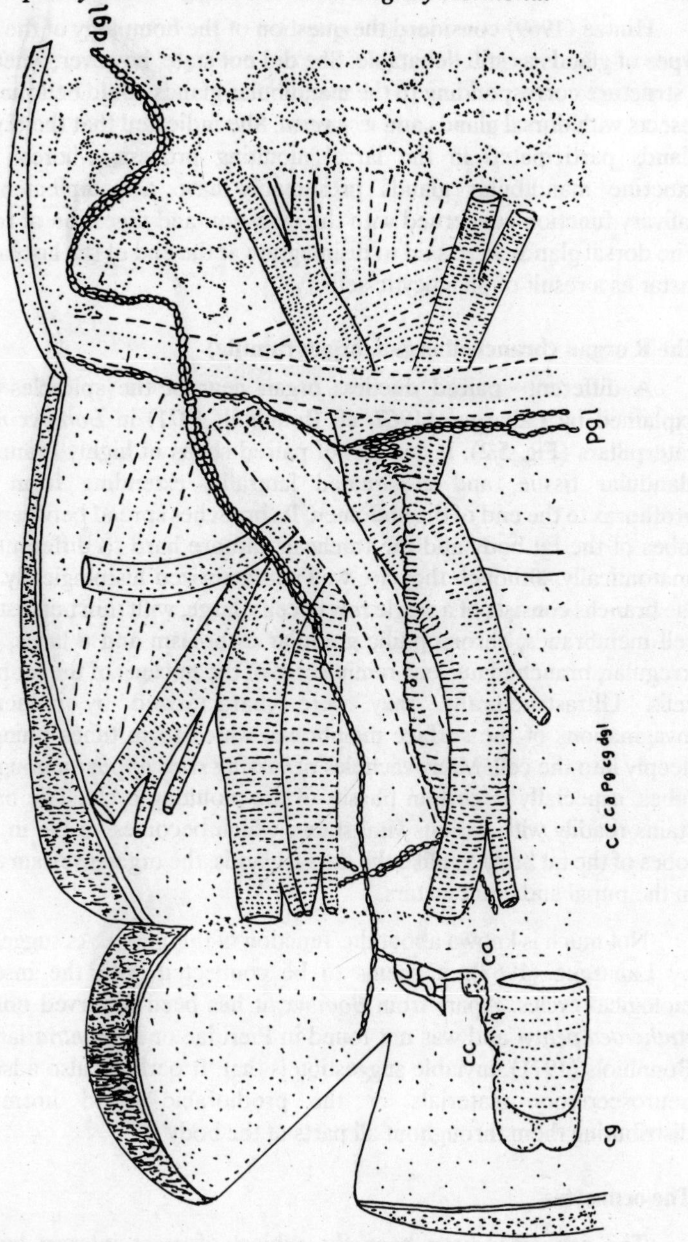

Fig. 5.1. Localizaion of the dorsal thoracic gland (dgl) and prothoracic gland (pgl) in *Cerura vinula* caterpillar. *dgl* - dorsal thoracic gland, prothoracic gland,cg ∴ cerebral ganglion, *cc* ∴ corpora cardiaca, *ca* ∴ corpora allata.

Hintze (1969) considerd the question of the homology of the two types of glands as still debatable. She did not state, however, whether a structure corresponding to the mandibular glands could be found in insects with dorsal glands and *vice versa*. She indicated that the dorsal glands participated in the larval moulting process, whereas the exocrine mandibular glands presumably have a complementary salivary function concerned with the ingestion and digestion of food. The dorsal glands were said to disintegrate at the end of the last larval instar as a result of phagocytic activity.

The R organ (branched organ, organ ramifie)

A different paired ductless organ near of the spiracles was explained by Lahargue (1967, cf. Bounhoil, 1971) in *Bombyx mori* caterpillars (Fig. 5.2). It consists of paired strips of highly branched glandular tissue, and is situated laterally, extending from the prothorax to the end of the abdomen. Its branches spread between the lobes of the fat body and the tracheae and are hard to differentiate anatomically, although they are well characterized histologically. All the branchs consists of a single row of cells, large, with thin but distinct cell membranes, chromophilic granular cytoplasm and a large, flat, irregular, branched nucleus reminiscent of the nucleus of the fat body cells. Ultrastructurally, they were characterized by numerous invaginations of the surface membrane, which form tubes plunging deeply into the cell. Many vacuoles of varying sizes appear among the tubes, especially in certain phases of the moulting cycle. The organ stains readily with various vital stains, which becomes visible in the lobes of the fat body. Unlike the dorsal glands, the organ even survives in the pupal and adult instars.

Not much is known about the function of this gland. As suggested by Lahargue (1967), it seems to be connected with the insect's biological cycles. Apart from *Bombyx*, it has been observed only in *Antheraea pernyi* and was not found in Pieridae or *Lymantria* larvae. Bounhiols (1971) unviable suggestion is that 'It perhaps also adsorbs neurosecretory materials or the prothoracic gland hormone, distributing them throughout all parts of the body'.

The oenocytes

The oenocytes have been the subject of great interest by the researchers involved in the endocrine functions of insects. Inspite of the fact that many authors agree that the oenocytes are in some way connected with humoral activity their function is yet unclear. Their cycles of activity during successive instars in the larval development of the chironomid *Syndiamessa branicki*, has been studied in detail by

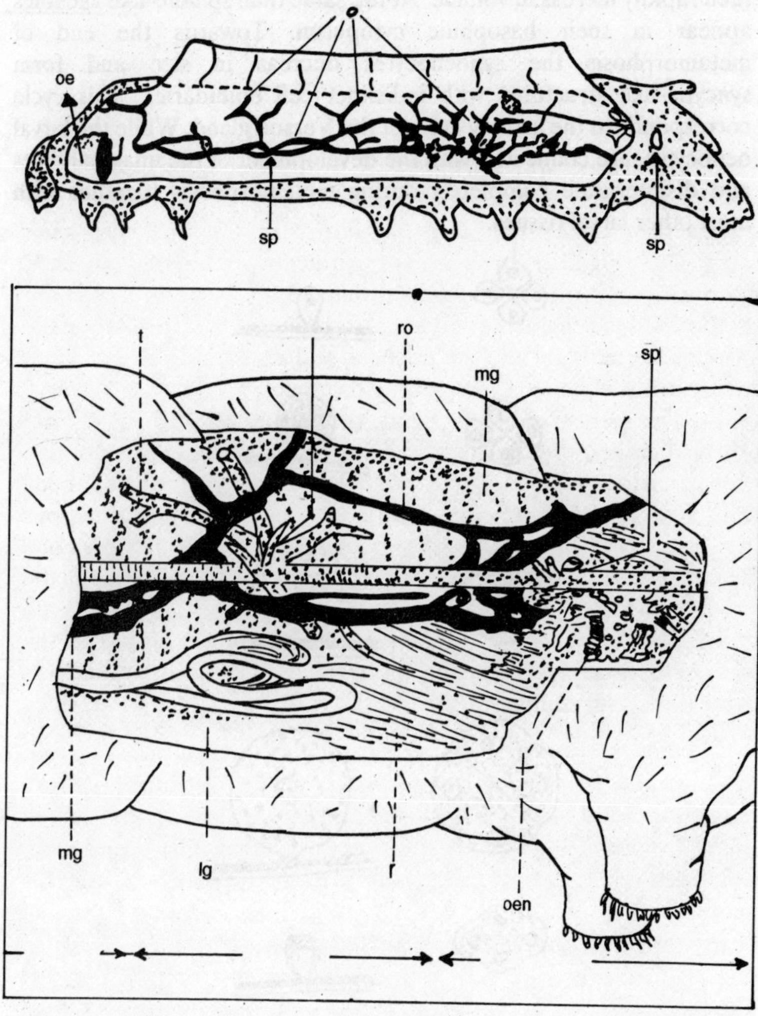

Fig. 5.2. 'Ramified organ' of *Bombyx mori* caterpillar. A, whole larva. B, one segment, *ro* – ramified organ, *oe* – oesophagus, *sp* – spiracles, *mg* – midgut, l – labial gland duct, *r* – labial gland reservoir, *oen* – oenocytes, *t* – tracheae.

Zavel 1953a). He observed that the cycle of activity of the imaginal oenocytes (synoenocytes) was closely connected with moulting process and with the development of the gonads, particularly in this species the synoenocytes do not appear until the last larval instar and then rapidly increas in volume. At the same time spindle-like vacuoles appear in their basophilic cytoplasm. Towards the end of metamorphosis the synoenocytes decreas in size and form syncytia-like structures with indistinct cell boundaries. This cycle corresponds to the activity cycle of the Verson gland. While the larval oenocytes, are connected with the development of the imaginal discs and disintergrate completely during metamorphosis together with most other larval tissues.

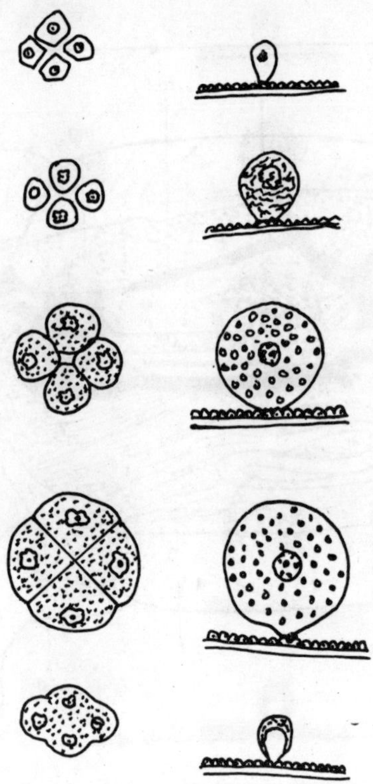

Fig. 5.3. Secretory cycle of oenocyte group (four cells) and a Verson gland in the epidermis of the chironomid *Syndiamessa branicki* in five consecutive periods of the last larval instar.

Regarding the larval oenocytes, many entomologists have stated that they produce a hormone which induces moulting. Actually in silkworms, the oenocytes are the first cells to show any activity during the moulting cycle, whereas they show no activity in a mutant strain which fails to moult. In the bug *Rhodnius*, however, the oenocytes do not reach their peak of activity until long after the moulting cycle has begun. In this species a specific relation between the oenocytes and the deposition of the cuticular layer of the epicuticle has been observed. Immediately before the deposition of this layer, the oenocytes become greatly enlarged and lobulate. At this time they stain deep grey with osmium tetroxide and stain spreads with fat stains. As the oenocytes discharge their contents and sub- sequently shrink in size, these straining reactions are shown successively by the epidermal cells and the newly formed cuticulin layer of the epicuticle. From these observations Wigglesworth (1949) suggested that the oenocytes should be regarded as epidermal cells which have become specialized for the production of the lipoproteins forming the cuticular layer and perhaps the wax layer in some other insects.

A connection between the imaginal oenocytes and the development of the gonads has been observed in several species. Definite changes are associated with the beginning of sexual maturity. It is probable that here also the oenocytes are involved in the production of lipoprotein forming part of the eggs or their shells. This is supported by the observation that oenocyte activity is much greater in the females of *Rhodnius prolixus* than in the males (Wigglesworth, 1933,1953a). This type of activity seems to be of a humoral nature (cf.p.397).

The secretory activity of oenocytes

The secretory activity of oenocytes was studied in detail by Schmidt (1961a, b) in *Formica polyctena* during postembryonic development Both larval and imaginal nocytes remain attached to the epidermis during the whole of the larval period. During the initial stage of the internal meta- morphosis in the pupa, they are looseded and accumulate at the sites where the imaginal organs develop. The larval oenocytes appear inside the midgut and are desisted when body pigmentation starts. Since the imaginal oenocytes and the trophocytes become fixed in the tissue at the same time, the haemolymph then become clear. The size of the oenocytes is determined by their secretory activity; the larval oenocytes attain their maximum size shortly before pupal moulting, the imaginal oenocytes before imaginal moulting. The oenocytes are asssumed to produce enzymes necessary during metamorphosis and are thus of primary importance in the formation of the imaginal organs. The imaginal oenocytes seem to have a function in the production of eggs. It was suggested that they were controlled by the corpora allata. There is no evidence which miTght indicate that they possess hormonal or excretory functions.

There has been lots of discussion on the contribution of oenocytes in the moulting process and the research has been stimulated by recent findings on their ultrastructure. Delachambre (1966), in an ultrastructural and histophysiological study, was the first to observe that they had a structure similar to that of mammalian steroid-producing cells. Like Wigglesworth (1933, 1948), however, he agreed that they participate in the formation of the epicuticle (and possibly of other cuticular layers) and assumed that they play a role in the biosynthesis of lipoproteins, the lipid fraction of which is rich in sterols.

While an entomologist named Locke (1969), in a very detailed and compre- hensive electron microscope study, concluded that 'the oenocytes, rather than the prothoracic glands, could be the source of ecdysone and the stimulus for moulting'. He infered that they interacted together with the brain and the prothoracic glands in inducing the moulting process and that they were actually the source of ecdysone. Two of his arguments are unacceptable firstly, that, judging by the structure of the prothoracic glands, it is unlikely that they produce steroid hormones (some authors observed smooth and tubular endoplasmic reticulum in them and secondly, that the more abundant structure appropriate for steroid synthesis in the oenocytes is, *per se*, evidence of ecdysone production. Deducing from the above arguement above, it seems possible that the amount of steroids needed for cuticular lipoproteins may be much greater than the amount required for MH. This does not necessarily exclude the theory that some of the oenocytes might function as an auxiliary MH-producing organ, however. At all events, the structure and phylogenesis of Locke's second type of oenocytes (the 'moult cycle oenocytes') demands more study.

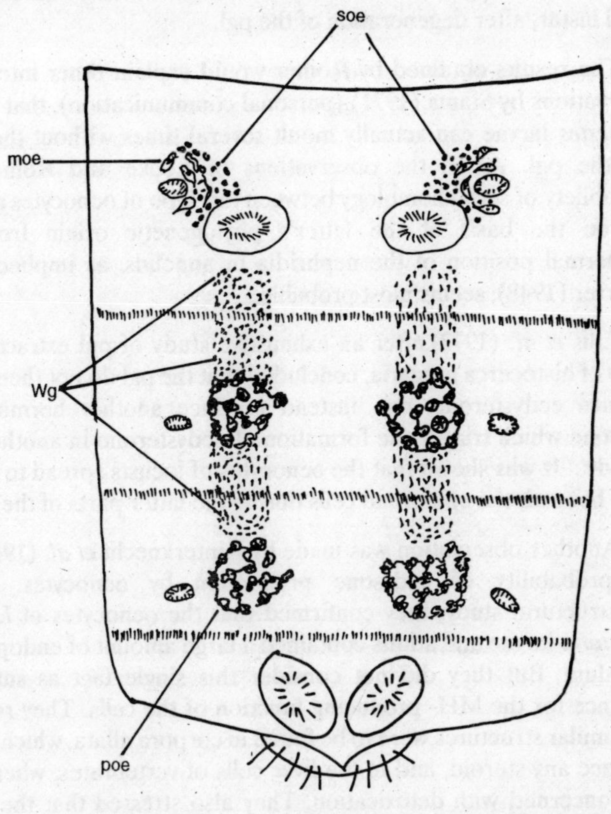

Fig. 5.4. Localization of three types of oenocytes in *Calpodes ethlius* larve. ,) *soe*—subdermal oenocytes, *moe*—moulting cycle oenocytes, *poe*— permanent oenocytes, *wg*—wax glands.

Following the lines of aforementioned arguements,, the experiments and observations of Romer (1971) on the oenocytes of the meal-worm *Tenebrio molitor* are very interesting. Romer tested oenocyte extracts obtained from 15 g of *Tenebrio* pupae in *Calliphora* and found them to be almost twice as active as prothoracic gland extracts prepared simultaneously from the same amount of prepupae and 4 to 10 times more active than the rest of the body. Oenocyte homogenates yielded similar results. The injection of ^{3}H-cholesterol showed strong incorporation into the oenocytes. Romer (1971) deduced that the oenocytes might be another source of MH (in addition to the pgl) and that they would be the only source in the pupal instar, after degeneration of the pgl.

The results obtained by Romer would explain other interesting observations by Slama (1971) (personal communication), that ligated *Dermestes* larvae can actually moult several times without the brain and the pgl. As in the observations of Locke and Romer, the probabilety of serial homology between this type of oenocytes and the pgl, on the basis of the latter's phylogenetic origin from the ectodermal position of the nephridia in annelids, as implied by B. Scharrer (1948), seems most probable.

Ellis *et al*. (1972),after an exhaustive study of pgl extracts from nymp of histocerca gregaria, concluded that the pgl do not themselves produce ecdysterone, but 'instead produce another hormone, or hormone which trigger the formation of ecdysterone in another part of body'. It was shown that the oenocytes of locusts spread to form a sheet beneath the epidermal cells during the latter parts of the instar.

Another observation was made by Rinterknecht *et al.* (1969) on the probability of ecdysone production by oenocytes. In an ultrastructural study, they confirmed that the oenocytes of *Locusta migratoria* larvae and adults contained a large amount of endoplasmic reticulum. But they did not consider this single fact as sufficient evidence for the MH- producing function of the cells. They recalled that similar structures were to be found in corpora allata, which do not produce any steroid, and in the liver cells of vertebrates, where they are concerned with detoxication. They also stressed that they were unanable to find a discernible difference in the function of the oenocytes in larvae an adults (although this may not apply to Locke's 'second type' of oenocytes) and they stressed the great variability in oenocytes of the same age. They suggested that it was far more likely that oenocytes have a multiple metabolic function.

In a thorough ultrastructural study of the moth *Calpodes ethlius* (Lepidoptera, fam. Hesperidae). Locke (1969) distinguished three different groups of oenocytes: (1) Permanent oenocytes, arranged in two paired groups below the wax glands; these remain enlarged throughout the intermoult periods and seem to have some relationship with the synthesis of wax precursors, as suggested by Kramer and Wigglesworth (1950) in *Periplaneta*. (2) Moult cycle oenocytes (see below), which Locke accepts to be concerned with ecdysone production. (3) Subdermal oenocytes which are claimed to be differentiated afresh from epidermal cells shortly before the pupal moult and most probably correspond to Zavel's 'imaginal oenocytes'

(.The indication which confirmed the theory that MH is produced by the pgl was recently presented by Agui *et al.* (1972), who showed that active pgl of *Momestra brassicae* were capable of inducing moult- ing in a piece of integument of rice stemborer (*Chilo suppressalis) in vitro*. Extraction of the active principle from the culture medium showed it to be acetone-soluble and thermostable. This conclusion was aslo congirmed quite independently, by Postlehwaith and Schneiderman (1970) who showed that implantation of the first leg disc of *Drosophila melanogaster* into a 3-day-old fertilized adult female of the same species was followed by its more or less complete differentiation, together with cuticle and bristles, and by Kambysellis and Williams and Kafatos (1971), who demonstrated that spermatogenesis in intact testes of the silkworm *Samia cynthia* can be induced equally well by adding ecdysone to the culture as by explanted active pgl. Inactive pgl are also active if an active brain of the same species is simultaneously.

The problem of sex hormones in insects

The initial efforts to demonstrate the presence of insect hormones were concerned with the gonads. The reason for this was probably that the first and most striking endocrine effects in vertebrates had been found to be those of the morphogenetically active sex hormone. The failure which followed such research with insects was probably the main reason why the first discoveries of hormones in insects were so delayed in comparison with hormones in vertebrates. Today it is considered proved that there are no subs- tances in insects equivalent to the morphogenetic sex hormones which influences the secondary sexual character of vertebrates. One of the most persuasive cause for such belief is the existence in insects of several types of gynandromorphs (mosaic, transverse and longi- tudina`), i.e. specimens which combine the secondary sexual characters of males and females. These type of creatures are not found among vertebrates. As all parts of the insect body are permeated by one body fluid (haemolymph), they would definitely all come into contact with any substances discharged into it. Were any of these substances were to influence differentiation towards either male or female, all parts of the body would surely be affected in the same way. Therefore there is no possibility of sharp transition between male and female parts.

Such assumption is supported by the nearly all negative results produced by work with experimental catration. The few exceptions for which the results were inconclusive are discussed in more detail below. Numerous experiments carried out by a number of researchers in a variety of insect groups, involving the removal of the gonads, have not resulted in any appreciable changes in the secondary sexual characters). Some entomologists refer to the results of experiments by Prell (1915) with the moth *Cosmotriche potatoria* as demonstrating an exceptionally positive effect of castration. Prell found that the castration of male caterpillars produced the pigment pattern characteristics of females in some of the resulting moths. In addition, females with the ovaries removed developed to moths with green haemolymph, which is colourless in normal specimens. But the same kind of experiments by the same author with *Lasiocampa quercus*, did not yeild the same results

Rejecting the probability of the existence in the *Cosmotriche* material of intersexes, which are quite common and would appear independently of the experimental treatment, the most probable explanation of Prell's results appears to be an artifact produced by the experimental treatment. Such treatment could have intervened with the factor correlating the reactions rates of the various developmental processes. This factor probably affects the development of secondary sexual characters in this species by influencing a single morphogenetic process in dissimilar manner

The 'androgenic hormone' of glow-worms

The presence of a humoral factor manufactured by a perticular cell in the follicles of the growing testis was demonstrated and analysed in detail by Naisse (1968a, b) in the species *Lantpyris noctiluca*. In these species, the gonads are differentiated to male and female organs during the IIIrd larval instar. When the factor in question in administered to female larvae after the next ecdysis, i.e. during the IVth instar, by the implantation of a paior of testes, the females are transformed to more or less normal males with both primary and secondary sex characters. The source of the hormone are particular cells in the apical part of each testicular follicle, which are differentiated during the IVth and Vth larval instars. The cells evolve secretory activity in this period, start to degenerate towards the end of the Vth (last) instar and then disappear, to be replaced by growing cysts in which spermio- genesis takes place. The cells dissapear completely during the pupal period. The female larvae undergo one additional larval moult and, after a pupal instar, are transformed to the neotenic (wingless) females, which is the distinguishing feature of this group of insects.

The entomologist later studied the neurosecretory system of this species and distinguished, in the pars intercerebralis, a special type of small-granuled neurosecretory cell with different activity in males and females. In males, the secretory activity of the cells started earlier, prior to sexual differentiation (IVth larval instar), while in females the granules in the cells appeared at a later stage and were less abundant and less conspicuous.

These type of changes were not seen in the other two types of nsc in the pars intercerebralis. The pupal and adult ca were virtually inactive in males, but in females they remained active. The pgl degenerated during the pupal period in males, but in females they remained partly till the adult period.

In another research paper, Naisse (1966b) summed up her experimental evidence. Bilateral extirpation of the cc and ca in the IIIrd instar transformed a male larva to a female. Similar experiment in any female larva, or in older male larvae,brought about no change. The author suggests that the neurosecretion of the small-granuled nsc produced in association with male sexual differentiation induces such differen- tiation via androgenic hormone stimulation. However, another, perhaps more feasible possibility should first be excluded. This is that the 'androgenic' hormone merely inhibits ca activity and that its pro- duction is determined by the neurohormone from the small-granuled cells. Another possibility, which as considered till now is that the activated or implanted gonads simply consume JH, thereby eliminating it from the haemolymph and reducing its concentration. Therefore, as in other insects, there would be no need to ascribe this particular phenomenon to a special andro- genic hormone.

Effects of parasites

The undeniable proof of the existence of special sex hormones are the numerous cases of morphological effects resulting from the stylopization, mermithization and internal para- sitism by larvae of entomophagous insects. For instance, stylopized females of solitary bees of the genera *Andrena* and *Halictus* lack various secondary sexual characters, such as the pollen brushes and pollen baskets on the third pair of legs. Also, the shape and pattern of the clypeus resembles the male condition to a greater or lesser extent. Similarly, parasitized males resemble females in some particulars. But, these phenomena can be easily clarified, without supposing a special sex hormone; the changes are the effect of the reduction in the amount of the food available to the host induced by the parasite. This explanation is supported by the observations of Schwanvich (1949) that stylopization can only affect the normal course of morphogenesis if the host larva is unable to replace the quantity of food consumed by the parasite. This is exactly what happens in solitary bees and wasps, but not in the social species. In solitary Hymenoptera, e.g. the genera *Andrena* and *Halictus* men- tioned above, the larva is enclosed in a mud cell and is supplied with a some amount of pollen and nectar by the mother bee. The cell is then closed, so that the larva is restricted to a particular amount of food, much of which may subsequently be used by the parasite. In contrast, the larvae of the social bees and wasps are fed directly by mouth-to-mouth feeding from the workers. As the amount of feeding depends upon the needs of the larvae throughout their development, no deficiency in parasitized individuals can occur. In fact, no morpho- logical effects have so far been observed in stylopized specimens of the social species, although the occurrence of stylopization is quite recurrent.

Many types of outcome of the internal parasitism of ant females lead to the same conclusion. With ants, the caste differences between the reproductive females (queens) and the sterile workers or soldiers are modified by parasitism the same manner. Contemporary research has proved that the differences between castes are influenced by differences in the quality and quantity of food during particular periods. Females in *Lasius niger* parasitized by the nematode *Mermis* (known as mermithogynes) have their abdomens almost entirely filled by the worm, and are obviously smaller and have a narrower thorax and shorter wings than normal females. Similarly, the mermithized workers of the Texas ant *Pheidole commutata* described by Wheeler,

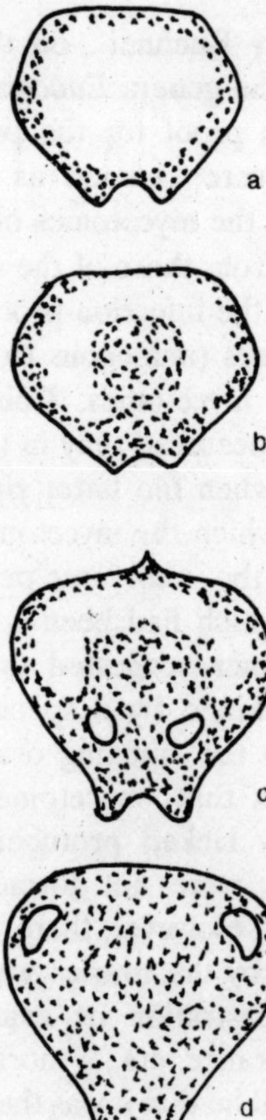

Fig. 5.5. The effect of stylopization on the form, size and pattern of the clypeus in the wasp *Odynerus perennis*: (a) clypeus of normal male, (b) stylopized male; (c) stylopized female, (d) normal female..

approach the female condition in having developed ocelli and a broader thorax. It is nor certain whether the assumption of the fact that the parasite induces female (queen) characters in the worker larva, is correct. It is much more probable that the queen larvae are so strongly affected by the reduced nutrition that their adult appearance resembles that of workers more than that of reproductive females. In these instances the secondary sex and caste characters are controlled by a morphogenetical mechanism which is based on the quantity or quality of the food consumed by the insect during a critical period as a larva. Such a mechanism is therefore easily affected by parasitism.

Obervations made by Buchner on the sexual differences in female leaf- hoppers of the genera *Euacanthus* and *Solenocephalus* similarly considered as a proof for the presence of a special sex hormone, though they were claimed as such by Koller (1938). Bruchner illustrated that the mycetomes of the female leaf-hoppers in these genera differed from those of the males by a special protuberance, which he called the infection-protuberance. It was through this that the infective stages (analogous to spores) of the symbiotic micro-organism left the mycetomes. Coincident with the protuberance, infective stages occurred only in the females. These stages penetrate into the eggs when the latter rippen and form a special globular structure from which the mycetomes of the new individual develop in the course of the embryonic period. The study on three females, the ovaries of which had been completely destroyed by a dipterous parasite, Buchner suggested that the formation of the protuberance was determined by a special substance or perhaps several substances which the ripening ovaries discharged into the blood. Inspite of the fact that mycetomes of these females were normally developed they lacked protuberances and the infective stages of the symbionts were not formed; thus the mycetomes resembled those of males. However, there was no humoral effect of any substance produced by the ovaries was shown. Even if it were possible to show that castration or ovary implantation affected formation of the protuberance, etc., a more probable explanation of the phenomena would still be that it was the cause of variations in the internal medium (the requirements of the quickly growing ovaries altering the nutrient balance otherwise available to the mycetomes).

Substances affecting the ovarian cycle in insects

Ivanov and Mestcherskaya (1935) Exhaustive study carried out by on the development of the ovaries, egg-laying and the shedding of the oothecae in females of the cockroaches *Blattella germanica* and *Blatta orientals*. They also examined the equivalent periods in the ovarian cycle of the migratory locust (*Locusta migratoria*). They discovered that, besides the progressive histological changes associated with egg ripening, rapid physiological changes occur in the ovaries of adult females. These changes could be reversed and were primarily recognized by a change in the permeability of the follicular epithelium covering the individual oocytes.

The ovaries, which have not yet attained maturity, as found in the nymphs of *Blattella germanica*, are characterized by the impermeability of the follicular epithelium to various vital stains such as neutral-red and haemoglobin, to some salts and to hydrogen peroxide. However, at a later stage, permeability to such vital stains is markedly increased and penetration occurs into the oocytes, and even the cytoplasmic granules are stained. This period remains until the oothecae is formed, when the per- meability of the follicular epithelium again decreases. The increase in permeability does not reoccur until after the oothecae has been shed when, correlated with its appearance, more eggs ripen (Fig. 5.6).

The substance from the fat bodies which affect ripening of the ovaries

The same two entomolosists (Ivanov and Mestcherskaya, 1935) discovered that ovarian maturity and its associated permeability characteristics could be induced artificially in immature nymphal ovaries by transplanting them into the body cavity of a female with maturing eggs. *In vitra* experiments demonstrated that this was in fact an effect of humoral factor(s) carried in the haemolymph. By testing extracts of various organs from mature females it was illustrated that, besides the haemolymph, extracts of the fat body were also effective. The researchers concluded that the fat body produces a hormone which induces ovarian maturation. But simultaneously they

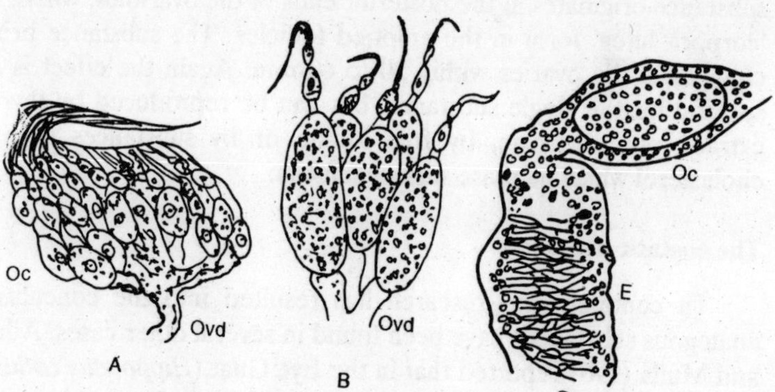

Fig. 5.6. *Blatella germanica*. A, immature ovary of nymph. B, ovary of a sexually mature female. C, longitudinal section of an ovariole of a female with ootheca; *Oc* – oocyte; *F* – empty follicle with corpus leuteum; *Ovd* – oviducts.

demonstrated that the effect is non-specific, extracts of vertebrate pituitaries having a similar effect when injected into the body cavity of a female with immature ovaries. The fact that maturation of the ovaries depends on the increased permeability of the follicular epithelium led the authors to suppose that a similar effect might be produced by various sub- stances which lower surface tension. Saponin and sodium tauro- cholate were tested and proved active, both *in vitro* and when drunk in an aqueous solution by females with immature ovaries. Unlike extracts of the fat body and pituitary, however, these substances did not induce normal maturation of the eggs. Although growth of the ovaries was hastened, maturation halted at a certain point so that no mature eggs ready for oviposition were acquried

The substance from the 'corpus luteum' inhibiting maturation of the ovaries

The momentary return to immature conditions is the distinguishing feature of those species in which the female carries the eggs in an ootheca until they hatch, including the cockroaches studied by Ivanov and Mest- cherskaya. In such species maturation of additional eggs during this period would be useless and might even be harmful. Transplantation of a mature ovary into the body cavity of a female with an ootheca has shown that here also the return to immaturity is due to a substance carried in the haemolymph. The substance originates at the posterior ends of the ovarioles, where the 'corpora lutea' form in the emptied follicles. The substance brings changes in the ovaries within 30 to 60 min. Again the effect is not specific for the single substance, but can be reproduced by thyroid extract, pure thyroxin, by free iodine or by substances such as cholesterol which increase surface tension.

The oostatic hormone

Th contemporary research has resulted into the conculusion analogous substances have been found in several other cases. Adams and Mulla (1967 reported that in the Eye Gnat (*Hippolatus collusor*) the oocytes of the second ovarian cycle did not develop beyond a certain stage (4) until mature first cycle eggs had been laid. An oostatic hormone was created in the ovaries of the house fly (*Musca domestic*) when the developing eggs were between stages 4 and 10. It could be derived from mature females from stage 2 onwards. It was soluble in 95 percent ethanol, in cloroform- methanol (2:1), in 20 per cent glycerol and in water. If injected into female flies within 24 hours after ecdysis it inhibited ovarian development. A semi-purified extract of the hormone was obtained from mature female homogenate.

The above obervations dustinctly indicate the probability the permeability changes in the follicular epithelium, which are the outcome of certain far less specific effects of various low and high molecular weight chemicals generally found in the organism.

The bursa copulatrix factor

Mated Hyalophora cecro-pia females lay far more eggs than vi rgins. If the bursa copulatrix of a mated female is implanted into virgin females, they lay the same numbers of eggs. The implantation of bursa copulatrix from a young, immature female has no effect in the same manner, the haemolymph of mated females is likewise active, but that of virgins is not (Riddiford and Ashenhurst, 1973). The authors conclude that a blood borne factor is produced by the bursa copulatrix after its contact with sperm or some element derivd from the male reproductive system.

Other alleged source of hormones in insects

The revelation of a new hormone called *proctodone,* developed by special epithelial cells of the anterior portion of the hindgut, was reported by Beck and Alexender (1964a, b) . Ligaturing experiments performed on diapausing last instar caterpillars of Orstrinia (= Pyrausta) *nubilalis* show that diapause development is dependent on a blood-borve factor originating in the seventh and eighth abdominal segments. The authors conjecture that proctodone stimulates the neuro- endocrine system to produce activation hone. Under a series of experimentally accelerated diapause development, the action of this factor seems to be required for about two days, the source of the brain hormone then being necessary for about five days, and the prothoracic system for about six days, so that to complete diapause and prepupal postdiapause development. The hormone was presumably produced periodically during the light hours of the 24-hour cycle. It was taken for granted that it affected in non-diapause growth, photoperiodism, several forms of diapause and polymorphism. Its possible connection with the factor supposed by Davey (1962) to be produced by the epithelial cells of the anterior intestine under the influence of a corpus cardiacum factor, which stimulates muscle contractions in the hindgut was discussed. It was also hinted that it was found in the non-photoperiodic and non-diapausing species *Galleria mellonella*. But even in this case, it is expected to activate the endocrine function of the brain. A number of questions still remain to be answered, however, before the mechanism can be accepted as general in insects. The transplantation of a supposedly active hindgut into diapausing *Ostrinia* larvae has so far not yielded conclusive results. It would also differentiate with the generally accepted assumption that neurohormones like AH are primary hormones, i.e. that they are produced in response to external stimuli which, in turn, they transform to humoral stimuli activating the production of secondary, glandular hormones.

Previous research pappers stated that the so-called Verson glands, more recently and correctly called dermal glands as in Verson's original terminology had a secretory function. Later work has shown that there are insufficient grounds for this view. Their role in the formation of the cement layer and, at least partly, of the wax layer of the epicuticle has now been shown in a number of species. Its function was first implied by Wigglesworth (1934,)

There are numerious cases where a secretory function of various tissues has been suspected. But none of them grant us any clear indication. Mokia (1941), for example, suggested the existence of a so-called egg-laying substance originating after fertilization in adult female silkworms. Haemolymph trans- fusions from a fertilized female into a virgin induced egg-laying in the latter within the first day of adult life, whereas generally it does not begin until several days after the imaginal moult. The myotropic effects of the neurohormones C and D mentioned earlier must also be taken into account in this connection.

Almost the same kind of suggestion was made by is Yakhontov's (1945), who stated that sexual maturity and the flying instinct in some beetles are humorally connected. There is also a regulating factor in the timing of the degeneration of the indirect flight musculature of female ants concerned with fertilization and egg-laying.

The effect of cyclical 3', 5'–AMP on the tanning of ligated *Calliphora erythrocephala* puparia was studied in detail by Knorre *et al*. (1972). Deleurance and Charpin (1971) studied a specific suboesophageal body with the characters of a ductless gland and suggested that it had a bearing on nutrition.

There are also many partially known substances of a presumably neurohormonal character, such as the eclosion hormòne, discussed in Chapter 7. Undoubtedly, there will be further research which would reveal a number of further humoral relationships in insects.

6

Hormones and Diapause

The subject which has greatly interested most of the entomologist, is *diapause.* It may be defined as a temporary interruption of development in a specific stage of ontogenesis which, in a particular species, would occur even if the insects were kept under optimum conditions. A span of opposing conditions of a definite character is often necessary before development can recommence. The opposite of diapause is quiescence, where inhibition of development is the immediate consequence of adverse conditions, at the end of which development is instantly renewed. It has been mentioned that the beginning of a 'true' diapause is independent of current environmental conditions; this does not mean that environmental conditions play no part at any point in the development of a given species in determining the occurrence of diapause. The occurrence of diapause in the most unfavorable conditions for the development of the species, for example the low temperatures of winter or summer drought, and its adaptive value for the species distinctively demonstrate the role played by environment in the evolution of a given species. Apart from this common phylogenetic adaptation, in so-called 'facultative' diapause, the ambient conditions during earlier stages of the life-history of the species concerned have been found to decide the occurance of the genetically conditioned 'facultative' diapause.

More information on the principle of diapause led to the discovery of two types of deviations from the above 'classic' definition.

The first includes cases in which a state corresponding to 'true' diapause may be induced within a long phase of ontogenesis, as in the IIIrd and VIIth larval instar of *Dendrolimus pini* , or during the larval development of *Lucilia*, when any opposing conditions such as unduly dry or moist food and wrong temperature induce arrested growth). This state is merged with and determined by various hibernating habits, as in *Eurygaster, Leptinotarsa* or Coccinellidae. In the second type, typical diapause can be disturb not only by adverse conditions but also artificially, e.g. by the action of long photoperiods, as in *Pyrrhocoris*. This can be easily confirmed by the recent discovery of the neurosecretory mechanism of diapause. As early as 1936, Wigglesworth showed that removal of the brain (by decapitation) at any point of postembryonic development induced in *Rhodnius prolixus* a state corresponding to diapause in every respect. In fact, all transitions between the two extremes, i.e. true diapause and quiescence, are found in various species, but those of the latter type are much more abundant, common, because of high selective value.

Diapause development

The span of adverse conditions (coldness) necessary for the recommencement of development varies with the species from two or three weeks to several (4 to 5) months. This period of arrested morphological development does not involve a complete rest in the physiological sense of the word; it is some what the time required for specific physiological changes to occur, which make it possible for the insect organism to restart development. It is therefore called the period of diapause development (cf. Andre- wartha, 1952).

The diapause development of a particular species has a definite minimum length at certain temperature. In most species where this aspect has so far been studied, the maximum temperature for diapause development is around 5°C. With both decrease or increase of temperature there is a remarkable increase in the length of diapause development, ranging from several weeks to several years as the temperature approaches the 'living' optimum (20 to 25°C), while development practically stops at temperatures below 0°C.

The stage at which diapause occurs

The occurance of diapause is generally at the most varied stages of ontogenetic development in different insects. For any one species, however, it is joined with a particular stage of development. There are many cases of embryonic diapause occurring at early embryonic stages (as in *Orgyia antiqua* or *Bombyx mori*), near the middle of embryonic development (as in *Locusta migratoria*, other Orthoptera and some of the Lepidoptera), or at the end of embryogenesis in a fully developed larva prior to hatching (as in *Lymantria dispar*). In the same manner, diapause can occur in any instar of larval development, for example the first instar in *Cacoecia fumiferana*, the second in *Euproctis chrysorrhoea*, the third in *Gastropacha quercifolia*, the penultimate in *Gryllus campestris* or the last larval instar in *Macrothylacia rubi*. Very often diapause occurs in the pupal instar, as in Satiurnia pyri, Hyphantria cunea, etc. Imaginal diapause mainly in-volves those parts of the body which continue to develop during adult life, primarily the ovaries f female insects (e.g. *Anopheles maculipennis* or *Pyrrhocoris opterus*). There are rave cases when diapause occurs at two separate stages within one generation (e.g. in *Dendrolimus pini* or *Cephalcia abietis*; see later).

Obligatory or facultative diapause

In few insects, diapause manifests itself in each generations with out any interference of the environmental conditions the animals experience. In others, it occurs only in some generations, when specific environ- mental conditions prevail or have prevailed at an earlier stage. The former type mainly occurs in insects with a long period of post- embryonic development lasting one or more years.When it happens in each generation, the diapause is termed obligatory. The second type occurs in insects which have more than one generations a year, and appears at the appropriate stage only in those generations which bridge the period of unsuitable conditions, e.g. the overwintering generations. In such instance diapause is facultative and is dependent on the appropriate conditions in the period of its determination. From the genetic point of view, either type of development (with or without diapause) has the character of a hereditary modification, as known among the castes of the social insects, for example, or in many plants, e.g. the dandelion (*Taraxacum officinale*). Special mechanisms are usually developed in some of the stages occuring before diapause when it is probably decided by specific conditions of, for example, photopreod and temperature. Whereas obligatory diapause is mainly associated with univoltine species, facultative diapause occurs in polyvoltine ones, i.e. bivoltine, trivoltine, tetravoltin or multivoltine. Alterations between obligatory and facultative diapause are known also where development can continue without a break for many generations under suitable conditions, but may easily be interrupted when unsuit- able conditions appear. When induced in such species, the delay in development has the character of a true diapause, but may easily be broken by specific environmental conditions, e.g. long photoperiod, and thus occupies an intermediate position between diapause and quiescence.This kind of evolution is discernible in lucilia erythrocephala and *Dendrolimus pini* .

THE PHYSIOLOGY OF DIAPAUSE

The significant feature of diapause is the inhibition of growth, is connected with many biochemical and physiological changes in the insect organism. These depict the adaptive nature of diapause, as they increases the ability of the insect to survive the adverse conditions (winter, drought). Some physiological characteristics are similar in all types of diapause:

Metabolism

Overall metabolism. The beginning of many types of diapause is correlated with a marked decrease in oxygen consumption. The respiration curve fails to a characteristic very low level— about one-tenth of the normal oxygen consumption of the particular stage—and this level is maintained practically throughout diapause. Diapause is generally associated with the bottom of the oxygen consumption curve for non-diapausing insects.

Many entomolgists have studied the respiratory metabolism during diapause and the influence thereon of the metamorphosis hormones. The oxygen consumption in diapausing adult females of *Pyrrhocoris apterus* was found to be similar with that occurring in normal (non-diapausing) specimens after removal of both ca and cc. It also corresponded with that in normal untreated males, whereas it was definitely higher in the females which had only been allatectomized. Typical low diapause respiration with a low water content and an increased amount of fat was found by Tombes (1964) in the aestivating curculionid, *Hypera postica* (Gyll). DNA synthesis during metamorphosis and diapause was examined by the incorporation of tritiated thymidine in *cecropia* and *cynthia* silkworms. In diapausing pupae, the incorporation of thymidine was limited to spermatogonia, haemocytes, and a few other cells. In the diapausing epidermal cells, DNA synthesis appeared only after a large epidermal injury and was limited to the immediate vicinity of the wound. However, a generalized incorporation was observed after the termination of diapause at the beginning of adult development. This is assumed to be the only proof for differentiating between the increase in metabolism of normally developing specimens and those which have been affected by injury (Bowers and Williams, 1964).

The respiratory enzymes. Many researches have demonstrated a series of changes in the activity of the respiratory enzymes. From the observed increase in the resistance to HCN and CO as well as a resistance (seventy times higher) to organic poisons which block the cytochrome system (i.e. pilocarpin and diphtheria toxin), it was concluded that the cytochrome system is inactivated during diapause and replaced by a particular cytochrome, b_5. But, further experimentation has led to a different conclusion. Investigations on the resistance to cyanide and CO of the heart = beat of diapausing *cecropia* pupae, which continues for $5\frac{1}{2}$ hours in the complete absence of oxygen, showed a large surplus of cytochrome oxidase in comparison with cytochrome c, producing the observed greater resistance. Kurland and Schneiderman (1959) attempted to generalize from these results to other types of diapause, including embryonic diapause, on the basis of their studies on the respiration of diapausing pupae of several saturniid silkworms (*Cecropia, Cynthia, Promethea, Polyphemus*).

They assumed that cytochrome c would be the factor constraining respiration during diapause, and that changes in respiratory metabolism would be the result of the lack of this factor as a direct or indirect consequence of MH.*. The distribution of cytochrome oxidase and succinic oxydase in adults of the Colorado beetle (*Leptinotarsa decemlineata*) during the period of the diapause period was studied comprehensively detail by Ushatinskaya (1957b). The results of her work and those of other Soviet authors are included in her monograph on the cold- hardiness of insects. The physiology of diapause in sawflies was studied in detail by Siama (1959a-c) in a series of research papers. The physiology of imaginal diapause in two coccinellid species (*Semiadalia undecimnotata* Schneid., and *Coccinella septem- punctata* L.) was studied by Hodek and Cerkasov (1958, 1959).

The utilization and accumulation of reserves. A significant physiological feature of diapause is the reduced utilization of reserve materials (protein and fats) correlated with the reduced metabolism. Prebble (1941) showed that, in the sawfly *Gilpinia polytoma*, no more than about 2 per cent of the reserve materials (dry matter) was consumed during the whole winter period. Similar results have been obtained with other species. This is probably one of the most important selective advantages of diapause. Where reduced metabolism occurs during suitable living conditions so that the diapausing insect is able to accept food, as in the so-called pre-diapause of the imaginal diapause, there is an accumulation of fats and other storing substances in the body of insect..

Water balance. An increase in the ratio of dry matter is another significant feature of diapausing insects making them more resistant to cold. Normal water balance is usually restored following the resump- tion of growth. The lack of required humidity may markedly prolong diapause in some cases. In summer diapause, as occurs in locust eggs, and less frequently in winter diapause, as in coding moth, diapause can be broken by immersing the insect in water. It is important to note that changes in the water balance as well as the afore-mentioned features of diapause seem to be the result of changes in the humoral balance of one particular endocrine factor, the AH, which controls, amongst other things, the water balance of the body.

Digestion during diapause. Many insects have unrestrained movement and are able to accept food during diapause. This is specially true of imaginal diapause, in so far as it occurs during suitable environmental conditions. It has been observed, however, that most insects incapable of utilizing the food, either for the development of the gonads or for restoring their fat-body reserves, so that they die of exhaustion within a few days. This can beunderstood as an AH-deficiency syndrome, now that its role in the activity of the digestive enzyme has been confirmed by Thomsen and Moller (1959) and by Strangways-Dixon (1960, 1961).

Since the last edition of this book there have been many new discoveries on various aspects of the physiological changes connected with diapause (e.g. Maslennikova, 1972). Although it is not fearble discuss them here, it should be mentioned that they should be regarded as consequences of the same principle changes in the insects neurohormonal balance.

Determination and photoperiodism

The determination of diapause

It has been affirmid that diapause coincides closely with the period of environmental conditions, adverse for a given species.

In multivoltine species, a special method is required to ensure such synchronization. Usually the insect has a hereditary struture which will respond with diapause to particular environmental conditions which normally precede diapause, such as a certain day-length (photoperiodism), or a change in temperature. Since diapause itself is closely associated to a particular stage of development, so is its period of determination. It is therefore hereditary. As indicated by Novak (1958b), the phylogenetic origin of this genetic mechanism can be explained as the result of selection for a tendency which perhaps occurs in all insects to a very small extent at certain stages of development. This tendency to respond to various inclement conditions, by a temporary break in development at some subsequent stage, may originally have no specific direction (e.g. for a long or short photoperiod) and only its value for the insect's survival under the particular environmental conditions may finally induce strengthening of another kind of response.

Therefore, it is proved in bivoltine races of *Bombyx mori* that optimal conditions during the embryonic period and during the first part of larval development (Ist and IInd instar) tend to produce egg diapause in the next generation. The same conditions during the IIIrd instar produce no effect on diapause. While, in the second half of larval development (IVthe and Vth instar) these same conditions (long photoperiod and higher temperature) tend to induce development without diapause and the converse also applies (cf. Emme, 1953). By specialization through selection for the early instar effect short-day photoperiodism results, while specilization on the latter effects result in long-day photoperiodism. The role of selection in the evolution of this type of dependence seems obvious from the markedly adaptive character of all the mechanisms.

The analysis of photoperiodism in connection to diapause has induced much interest since its importance was first stressed by Ernst Scharrer (1952, 1959) that it can now be regarded as a special branch of research on diapause. Basic work in this field has been carried out by Danilevski's team since 1945 . Even here, the neuroendocrine basis of all these relationships is in undoubtedly clear.

Different in degree and yet similar to photoperiodism is the action of temperature during the determining period of facultative diapause. It is interesting that the influence of temperature in a particular species works in the same direction as photoperiodism: in short-day species a low temperature has a positive effect in the same way as a short photoperiod while higher temperatures induce diapause; in long-day species the temperature and photoeperiodism act in the opposite response has a clear adaptive value, e.g. a low temperature together with a short photoperiod during embryonic development causes the deposition of non-diapausing eggs in the bivoltine race of *Bombyx mori*. It is widely accepted that the first spring generation of silkworms hatches in the early spring (March or April in China) when low temperatures and relatively short days are normal conditions. The females of this generation thus deposit non-diapausing eggs of the second generation which hatch in about June, when the days are long and the average temperature is high. In this way the females of the second generation are accustomed to lay diapausing eggs which are able to survive the induce development without diapause and higher humidity winter period, whereas a hatch in autumn would naturally caused death through cold and lack of food. The interaction of the effects of photoperiod and temperature when the conditions are varied experimentally is extermely complese (1949, Danilevske,)

Along with the effects of photoperiod and temperature, an influence of humidity on diapause has also been discovered in many species e.g. *Bombyx mori*. Again the effect of humidity is aligned with the influences of the other two factors; with short-day species, conditions of lower humidity induces development without diapause and higher humidity induces diapause. The effect of humidity in the summer type of diapause in insects from tropical and surtropial countries appears to be rather more involved, though and yet fully understood. The influence of food and its composition on the determination of diapause is also doubtful. Many results in the literature have proved the described cases to be unreliable when repeated with a controlled photoperiod. In other cases it is still unclear whether the observed effect is specific or merely very generally affects development, e.g. suboptimal nutrition might make the insect more susceptible to factors inducing diapause. There is, however, a definite case of an influence of want of food on diapause determination in blood-sucking insects. Here a special mechanism has developed which synchronizes development with periods of food abundance and induces a state corresponding to a diapause during periods of starvation. Shotrage of food also seems to be a stimulus inducing diapause in some plant-feeding insects, e.g. in *Diatraea lineolata*, a subtropical maizeborer.

Termination of diapause

As earlier mentioned, the degree of temperature during diapause development may produce a profound effect on the length of diapause. Never- theless, no change of temperature within ecological limits is able to break diapause at any given moment and re-induce development. But there are a number of other factors, including experimental treatments, which may be able to induce normal development in animals already determined for diapause. Perhaps the best known are the various treatments which suppress diapause in diapausing *Bombyx mori* eggs . It is important to note that most of these factors are only likely to break diapause during a definite, relatively short sensitive period; this is usually just before or at the beginning of diapause. For instance a short immersion of silkworm eggs examined for diapause in concentrated HCI can stop the process of diapause completely if performed during the first 48 hours after oviposition, i.e. before the eggs have their normal grey-blue colour. After this moment the acid treatment is completely ineffective. An effect on diapause similar to the HCI treatment can be obtained by increasing the temperature to about 46°C for to min in the same sensitive period. This procedure is such a potent stimulus that it can induce development, and hence artificial parthenogenesis, in unfertilized *Bombyx* eggs, the state which can thus be compared with true embryonic diapause.

The process of breaking diapause by increased photoperiod was studied by williams and Addisson (1964) in *Antherea pernyi*. They r ealized that long-day conditions (15 to 18 hours light promoted termination of the diapause whereas short-day conditions) (4 to 12 hours) prevented it. The experiments showed that the brain itself is the site of the action of light, most probably by nervous control of AH production. In this connection, the transparent window in the pupal integument immediately above the brain seems to be of primary importance, as shown by Shakbazov (1961). A photoperiodic acceleration of the termination of diapause was also found by McLeod and Beck (1963) in *Ostrinia (= Pyrausta) nubilalis*. A similar acceleration was observed following administration of large doses of ammonium acetate or other ammonium compounds. Syzergism in accelerating diapause development was found between the action of NH_4 ions and a long photoperiod. It has been shown that the effect of a long photoperiod in preventing diapause can be replaced by an additional short photophase (5 to 30 min) applied daily at about the middle of the dark period, even when combined with a very short photoperiod (e.g. 10 hours) and ecology of photoperiodism in insects was summarized.

Diapause in larvae of the fly *Lipara lucens* can be disturbed by a pin prick or by contact with a growing pin. Similar stimuli will also readily break the relatively feeble diapause in flies of the genus *Lucilia*. Diapause in eggs of the grasshopper *Melanoplus* can be broken at any time in its duration by immersing the eggs in xylo or other wax solvents. It has been demomstrated that effect of the xylol is to remove the waterproof wax layer (which covers the hydrophyle during diapause, so preventing water penetration) which is the fundamental cause of diapause. Removing both cuticular envelopes (the yellow and white cuticle) and immersing the eggs in insect-Ringer has the effect similar to the same experiment done by xylo It may be stated that the effect of strong mineral acids on breaking diapause, e.g. in silkworm eggs,lies in producing a particular temperature on the surface of the eggs, which is the main cause of the breaking of diapause.

The major cause for the diapause break in various examples of post- embryonic diapause by stimulation of the nervous system probably lies in the renewed secretory activity of the cerebral neurosecretory cells. The only treatment which is able to break postembryonic diapause at any stage of its development is to induce experimentally an active concentration of AH (by implantation of an active brain) or MH (by implantation of prothoracic glands or by the injection of ecdysone).

The mechanism of the effects of photoperiodism and other external factors

There is some proof concerning the way in which external stimuli are transmitted to the neurosecretory cells and thereby control their secretory activity. One of the earliest research papers relevant to this problem was written by Dixon (1949). He suggested, on the basis of his investigations on diapause in the oriental fruit-moth *Laspeyresia molesta*, that a series of hypothetical two-phase chemical reactions formed the basis of the response to photoperiod. One of these phases requires light, whereas the other needs darkness. Ushatinskaya (1961c) stated that in this context cytochrome oxidase might play a significant role Its inhibition by CO is prevented by light, this being a photosensitive reaction of a type which would be the basis of such a mechanism. The first hint on this question appears to come from a paper by van der Kloot (1955). He studied the sequence of changes in the electrical activity of the brain, i.e. the titre of cholinesterase (ChE), electrical activity (EA) and the amount of cholinergic substances in connection with the observed AH production in *H. cecropia*. He discoverd that all these changes are closely connected with AH secretion and, together with the latter, are closely connected with diapause. Thus electrical activity drops together with AH and cholin- esterase activity, remaining practically at zero throughout diapause with a steep rise afterwards. The cholinergic substances, however, follow an almost exactly opposite pattern (cf.fig. 57). The amount steadily increases during diapause development and reaction its maximum at the point when diapause ends. About this time, electrical activity reappears and rapidly increases together with the titre of the cholinesterase. It seems that the observed accumulation of cholinergic substances forming a substrate for cholinesterase is the direct internalcause of the effects of chilling, because it is optimal at low temperatures. It is confirmed that such a surplus of cholinergic sub- stances as a substrate for the renewal of cholinesterase activity is necessary for production of AH and the renewal of electrical activity. In the brain of the non-diapausing *Galleria mellonella*, the cholinesterase content and the amount of cholinergic substances is stable during metamorphosis. In the medial neurosecretory cells of *Bupalus piniarius*, neurosecretory material seems to be produced and correspondingly the spontaneous electrical activity of the brain persists during pupal diapause (Schoonhoven, 1962).

The existence of photoperiod by *Leptinotarsa decemlineata* has been investigated in a series of papers by de Wilde *et al*. (1959). The Colorado beetle is a typical long- day species, diapausing in a photoperiod of less than 14 hours of light. The threshold intensity of white light is less than 0.1 lux. Experiments aimed at localizing the site of perception have not as yet yielded clear-cut results. Removal of the eyes or interference with their function does not affect the photoperiodic reactions. The eyes are perhaps not the only sites of perception. Unlike Dixon and Ushatinskaya, de Wilde indicated a possible effect of feeding, perhaps connected with a rhythm of feeding activity. A review on the problem of photoperiodism in insects has been produced by Lees (1959). A monograph by Danilevski (1961) on the photoperiodism of insects is also significant.

THE HORMONAL PRINCIPLE OF DIAPAUSE

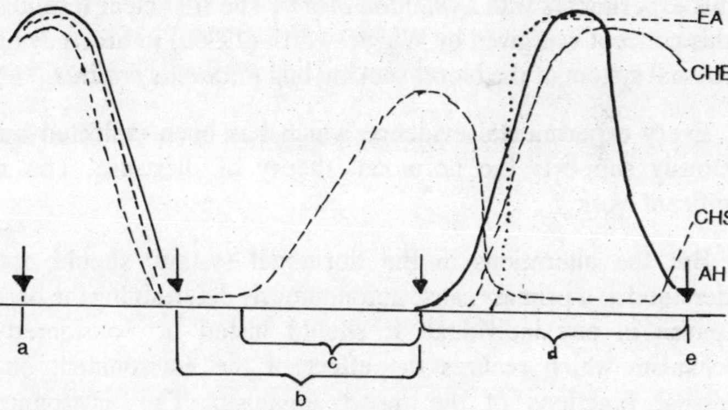

Fig. 6.1. The interrelationship in activation hormone production (AH), electrical activity of brain (EA), choline esterase content (CHE), cholinergic substances (CHS) in *Hyalophora cecropia* during the pupal instar. (a) beginning of cocoon spinning; (b) period of diapause; (c) period of low electrical activity; (d) period of development; (e) imaginal moulting (ecdysis).

A succession of many other theories has been formulated to explain the mechanism and the origin of diapause . The important ones can be divided into three groups.

(1) Diapause is caused by the accumulation of a specific inhibitory substance 'acting through the inhibition of the life processes' which is slowly decomposed during diapause development. This category also includes the concept of the general diapause hormone proposed by linton (1953). However, no clear proof of a substance with such characteristics has so far been acquired leaving the findings on the humoral conditioning of embryonic diapause in *Bombyx*.

(2) Diapause is the result of a decrease in the water content of the tissues caused among other things by the accumulation of reserve materials, primarily fats. This leads to the reduced metabolism and to the other features of diapause Diapause is, however, much more probably the cause rather than the effect of all these changes, as Andrewartha (1960) has stated, or perhaps the result of one and the same cause as these changes.

(3) Diapause is hormonally conditioned, either as a result of the absence of metamorphosis hormones indispensable for development, or, in the case of embryonic diapause, the result of the action of a special humoral factor. A suggestion that such is one role of the brain hormone can be found in Kopec's (1917, 1922) original papers based on his experiments with *Lymantria dispar*. The first clear formulation of this concept was given by Wigglesworth (1936a) in his study of the hormonal system of the blood-sucking bug *Rhodnius prolixus*.

Every experimental evidence which has been collected unambiguously supports the hormonal theory of diapause. The most significant work

But the alterations in the hormonal system, should not be understood as a primary cause autonomously determining the onset of diapause in any individual. It should insted be considered the mechanism which realizes the effect of the environment on the principal functions of the insect organism. The environmental conditions affect these functions either directly in the same generation by means of the nervous system and neurosecretory cells, or indirectly through previous phylogenetic stages, via the genetic mechanism under the perpetuae pressure of natural selection.

Three major types of diapause can be distinguished on the basis of their hormonal mechanisms: (i) Diapause caused by a lack of AH and (therefore) of MH. To this category belong all cases of larval and pupal diapause known so far and, doubtless, also late embryonic diapause. (ii) Diapause caused by lack of AH and JH; this mechanism is concerned in imaginal diapause. (iii) Diapause caused by the action of the neurosecretory factor produced by the suboesophageal ganglia of the female, and affecting the development of eggs. This category comprises early embryonic diapauses.

The diapause effected by a deficiency of the activation and moulting hormones is the most common type of diapause, and the type so far most thoroughly investigates. It consists of an arrest of growth and development due primarily to the absence of MH, itself a consequence of an inhibition of AH production. It was originally supposed by most authors that the lack of MH was the only internal cause of this type of diapause. Recent findings on the role of AH in digestion and water balance, however, make it very probable that, besides the indirect effect of AH in activating the prothoracic glands, there also persists a direct effect of the hormone, the lack of which causes an important part of the diapause syndrome.

Pupal diapause

The initial and comprehaensive research on of pupal diapause was carried out by William and his school, using the giant silkworm *Hyalophora (= platysamia) cecropia*. The main facts about this type of diapause may be briefly summarized as follows.

In this particular species, diapause begins at the same time as the ecdysis to the pupa. The remarkablely low metabolism of diapause begins somewhat later, however, at the trough of the U- shaped curve. Diapause metabolism, unlike that of developing insects, is highly resistant to the metabolic poisons HCN and CO and to toxins of bacterial origin such as diphteria toxin, and differs from normal metabolism in a variety of other feathers which disappear when development is renewed. Pupal diapause is often on ligatory as in *H. cecropia* and the Middle-Eastern saturniids (*Saturnia piri, S. Spini Eudia pavonia*), perhaps even be facultative as in the bivoltine *Antherea pernyi*.

Experiments with extirpation, transplantation, sectioning pupae, injection of extracts, etc. have clarified that the primary internal cause for the inhibition of development in pupal diapause lies in an interruption of the secretory activity of the neurosecretory cells of the pairs intercerebralis, and possibly includes entails a blocking of the passage of neurosecretory granules into the haemolymph. By implanting an active brain into the diapausing insect it is possible to break diapause at any stage of diapause development. The effect of chilling on the termination of diapause is especially concerned with the brain. This is demonstrated the fact that the implantation of the brain alone from a chilled specimen is able to break diapause in another unchilled pupa, whereas the chilling of a decerebrated pupa is ineffectual on diapause. Adding the AH source is, however, only effective when the host specimen retains its own prothoracic glands or when they are implanted at the same time. The implantation of one or more active brains into an isolated abdomen has no effect, whereas the implantation of active prothoracic glands breaks diapause and development continues. A similar impact is produced by injecting MH in the form of ecdysone prepared by the method used by Butenandt and Karlson. Little is known about which of the other known characteristics of diapause are the direct effects of the lack of MH, which are caused by the lack of AH, and which are perhaps an indirect consequence of growth inhibition. The results from the work on *H. cecropia* were similar to that obtained by other entomologists using diapausing pupae of other insects. Thus in Hyphantria cunea, Anterea pernyi and Eudia pavonia the implant- ation of either an active brain or prothoracic glands results in breaking diapause.

A exhaustive histological research on diapausing pupae of the moth *Mimas tiliae* was carried out by Highnam (1958a, b). He demonstrated that the neurosecretory cells of the brain are inactive in the diapausing pupa at room temperature, but that they produce intracellular material during the first 3 weeks of chilling (at 3°C). The material (granules) passes to the corpora cardiaca and towards the e nd of diapause development the neurosecretory cells again become inactive. A similar period of activity was observed in the corpora allata. The removal from specimens of voth corppora cardiaca and corpora allata, how- ever, had no effect on the duration of diapause. Such specimens had a notably thicker epidermis and their fat bodies contained a much smaller number of inclusions than the controls. The author supposes that the corpora allata play some role in maintaining diapause, perhaps by controlling the metabolism of the fat body in some way. No signs of secretory activity were observed while the subsequent development at 25C was taking place.

A number of research work has been published on the effects of AH and MH deficiency on metabolic enzyme during pupal diapause. Apart from there work mentioned above, Shapirio and Williams (1957a, b) studied the cytochrome system in various tissues of late larvae and pupae of *H. cecropia* using low temperature spectroscopy and a DU spectrophotometer. They report the presence in last instar larvae of a complete cytochrome system with moderate to high concentrations of cytochromes a, a_3, b, c and b_5, whereas in diapausing pupae no b and c were detectable by the methods used. When development continued after the termination of diapause, the activity of the original complete system reappeared. The explanation of the observations given by the researchers must, however, be compared with the later findings of Williams and Harvey (1958) and of Kurland and Schneiderman (1958), which have been earlier stated

It has beenstressed that a specific inhibitory factor, inactivated by cold, participates in the induction of a diapause state in diapause-determined *Pieris brassicae* pupae. It was shown, that such insects can recover from the chilling, and their pupae continue to develop even if the obtain is extirpated at the outset of diapause. Reactivation in the absence of brain AH might, however, be attributable to the amount of AH produced by the ventral nerve cord ganglia, but gernerally this does not acquire an active concentration.

Pupal diapause can take place at any point during the pupal instar. Such process in a given species is bound to the same phase of the pupal period and in most species it is genetically fixed. An example of a very late pupal diapause is the pronymph of the sawfly *Cephalcia abietis* (see below).

Larval diapause

We do not know much about the various examples of larval diapause. Based on what is known, it is apparently clear that larval diapause does not differ fundamentally from pupal diapause. It has been observed that in diapausing, penultimate instar larvae of *Gryllus campestris* the histology of the prothoracic glands shows reduced secretory activity and the brain appears to be responsible for these changes. The brain and prothoracic glands have also been found to control diapause in the larvae of *Sialis lutaria* . Larval diapause in ants of the genus *Myrmica* and its relation to caste polymorphism has been examined comprehensith by Brian (1959, 1961, cf. 1960).

Much of the research work dealing with diapause in larvae have been concerned with the diapause which occurs in sawflies. However, this is not a true larval diapause, but rather the diapause of an instar in metamorphosis occurring between the true larva and pupa. The existence of neurosecretory cells in the brain of the sawfly *Diprion pini* has been demonstrated by L'Helias (1953a-c). Church (1953, 1955) has shown that diapause in *Cephus cinctus* is caused by a temporary interruption of the neurosecretory activity of the larval brain, and that the renewal of this activity requires a certain period of chilling. The same author later showed that certain prothoracic glands described by him are necessary to terminate diapause. Two periods of diapause, separated by a period of limited interdiapause development, have been found in the pre-pupal stages, the eonymph and pronymph of another sawfly, *Cephalcia abietis* (Pamphilidae). Normally, this insect's span of development is two years. The first diapause period, which occurs at the eonymph stage, needs a considerably longer period of chilling (not less than four months) than the second period. This corresponds to the pre-pupal stage at an advanced stage of the moulting process. It generally occrus in the second winter, and is shorter under the same conditions as the first (eonymph) diapause, the preceding winter. The interdiapause development normally occurs in the summer months between the first an second hibernations. The pronymphal diapause can be broken by the implantation of an active brain. Ligaturing experiments have demonstrated a clear reliance on the brain and prothoracic centers in this sawfly, just as in other species.

Like pupal diapause, larval diapause can occurat any point during larval development. The IInd larval instar of *Euproctis chrysorrhoea*, which hibernates in specific common nests with over 200 caterpillars, can best stated as an example of a very early larval diapause.

Late embryonic diapause

There is no clear indication concerning the diapause which occurs in more or less completely formed larvae before hatch- ing, as is found in *Lymantria despair*. The fact that active neurosecretory cells and prothoracic glands can be demonstrated in embryos makes it clear that the hormonal mechanism at work at t*-his stage is the same as that in larval diapause, though this is not the case in younger embryos (see later).

Adult diapause

Imaginal diapause is caused by the clack of activation and juvenile hormones. A quite similar neurosecretory mechanism operates as in the aforementioned cases of true diapause, but with the difference that it is a lack of JH, not of MH, which is the main factor in adult insects. Moreover, the neurosecretory mechanism of JH is not fundamentally different from that of MH. The hormonal mechanism of imaginal diapause has been most thoroughly studied in Coleoptera, for example in *Dytiscus marginalis* and *Leptinotarsa decemlineata* and in Diptera, for example *Calliphra erythrocephala* and in the mosquitoes *Anopheles macullpennis Culex pipiens, C. molestus* and *Aedes aegypii* Simiar methds functions in the imaginal diapause of Heteroptera, as has been shown by Johnson (1958) with *Oncopeltus fasciatus* and by Slama (1962) with *Pyrrhocoris apurus*. Little is known about imaginal diapause in Lepidoptera, such as that which occurs in many common butterflies of moderate geographic zone like Nymphalidae or Pieridae which overwinter as adults; but a similar mechanism may well be accepted

Morphologically, imaginal diapause is only distinct in the interruption of growth of the ovaries and in the suppression of the functions of the accessory glands of both males and females. As has been shown by the experiments of E. Thomsen (1952) with *Calliphora erythrocephala*, this seems to involve a complex syndrome of both AH and JH deficiency. The implantation of corpora allata can cause the ripening of some of the eggs, but a quantitatively complete termination of diapause is only possible by the simultaneous action of both JH and AH. Besides this, the presence of AH seems indispensable for the corpora allata to function normally. The most important physiological effect of the lack of AH in diapausing adults is that on proteinase activity in the gut, which is probably also the cause, either directly or indirectly, of a variety of other physiological changes in species with imaginal diapause.

The connection of the corpora allata to imaginal diapause is clear from the necessity, first shown by Wigglesworth (1935, 1936) of JH for the ovarian follicle cells. Experimental evidence of the dependence of imaginal diapause on a temporary break in ca activity as suggested by Detinova (1945) for *Anopheles maculipennis*, was provided by Joly (1945) in his detailed study of *Dytiscus marginalis* and has since been confirmed by a number of other authors in various groups of insects which undergo imaginal diapause, e.g. by de Wilde *et al*. (1954-1961) in a series of papers on *Leptinotarsa decemlineata*; Hodek and Novak (1961) in *Eurygaster*; Larsen and Bodenstein (1959) in *Aed*es aegypti and other mosquito species, etc. Here, it is enough to say that the primary internal cause of imaginal diapause lies in the neurosecretory system of the pars intercerebralis, which appears to be part of the mechanism of action of external stimuli. The absence of AH, of course, produces a series of effects in the body, in addition to the inhibition of the ca, as shown by the papers of E. Thomsen and I. Moller (1959), Strangways-Dixon (1960) and others. This is because the state after allatectomy is not identical with the natural imaginal diapause and why diapause also occurs in males, the gonads of which are not dependent on the presence of JH. This has been clearly demonstrrated by Slama (1963) in his experiments on the effects of allatectomy, cardiacectomy and of the natural diapause on the oxygen consumption in *Pyrrhocoris apterus*.

A stage, often stated by entomologist as 'predia- pause' has been observed before hibernation in most examples of imaginal diapause in insects. It is accepted to be as a period of preparation for hibernation, during which various biochemical changes occur in the body to make the insect more resistant to cold and other adverse conditions, e.g. an increase of reserve materials (fat),lack of water content in the tissues, thus increasing their resistance to freezing, etc. Individuals of diapausing generations grow less quickly than those with continuous development, and related to this is the fact that they usually reach a greater weight and size during their periods- of development. Many authors have attempted to explain these changes as caused by an internal factor (conditioned by environmental con- ditions) which induces diapause. This supposition was made, e.g. in the so-called 'food mobilization' theory. There is, however, weighty evidence against this supposition. In the majority of insects undergoing imaginal diapause, diapause starts without such earlier changes in the organism, and appears to be fundamentally identical with 'prediapause'. Also, Ah or JH can terminate diapause following a preparative prediapause.

Easier definition of this stage is possible : there is no difference in physiological conditioning between 'prediapause' and the subsequent diapause; both being due to a deficiency of AH and JH. When this deficiency occurs within normal environmental conditions, feeding and the other functions of the body continue, but the ovaries and accessory glands are inactive. Food utilization is there- fore limited to energy and basal metabolism, with the result that surplus food accumulates in the form of reserve materials. Other changes are probably caused by the lack of AH activity on the gut proteinases. All the continuing processes interrupted during prediapause are arrested by the adverse conditions of winter. Prediapause is therefore undoubtedly true diapause, whereas what is regarded as proper diapause by the adherents of the 'food mobilization' theory is diapause of the ovary and accessory gland tissues combined with quiescence of all the other parts of the body, particularly the subsequent quiescence of the whole organism. The latter changes from normal quiescence, however, in that no hormones exist in the body at the time. Inhibition of the neurosecretory system is nevertheless over in that period. The results obtained by Hodek (1962) suggest, however, that in 'prediapause' the syn- dromes of AH and JH deficiency seem to be combined with inherited instincts for hibernating – such as the instinct for flight to hibernation sites in Coccinellidae, or in *Eurygaster integriceps*. This result is confirmed by recent findings on the direct effects of the metamorphosis hormones on imaginal instincts, as shown by Zdarek (1970) and others. More experimental evidence will be required, however, before the relationship between the release of these instincts and the onset of diapause can be explained It may consist in a parallel response to the same stimulus (e.g. a change of photoperiod, or one may excite the other (diapause the instincts, or vice versa, e.g. by keeping the insects in the dark, without food).

These conclusions on the nature of 'pre- diapause' corroborate with the fact that the occurrence of a prediapause is almost entirely limited to imaginal diapause and only affects a small part of the body while other tissues continue their functions. However, it should also be noted that in many insects diapause ends before the start of winter, the cold days and especially the cold nights of autumn providing insufficient chilling. By contrast, diapause may be terminated at the beginning winter simply by placing the insects in optimal conditions for further development. Inspite the state of such insects in winter corresponds in most respects to true diapause, it differs by corresponding with quiescence in that it may be broken by transferring the insects to laboratory conditions. As regards the other biochemical characteristics of imaginal diapause, apart from the direct effect of the lack of AH on proteinase activity, only the diapause of the ovarian tissues corresponds to a diapause caused by AH-MH deficiency but their state influences the tissues that develop the body.

Apart from water diapause, a summer dormancy in *Leptinotarsa decemlineata* has been described by Ushatinskaya (1961) as occurring in a proportion of the beetles surviving from the previous year. The beetles creep into the soil in the hot summer months (June, July) and remain there for about one month, after which they continue to feed. possibly summer diapause has the same hormonal mechanism as the winter one. In the autumn some of the 'old' beetles of the previous year enter another 'winter diapause' which differs to some extent from that of the first year.

Early embryonic diapause

Early embryonic diapause is connected with the diapause hormone (DH), a neuro- hormone of the suboesophageal ganglion.

The major distinctive feature between embryonic diapause (except late embryonic diapause mentioned earlier) and the two earlier categories is that in the present type determination usually takes place in the females of the previous generation, so that it must be assumed that it is transmitted via the eggs. The other important feature of this type of diapause is that it occurs at a stage of ontogeny before any of the metamorphosis hormones become active. The relatively small number of species in which the hormonal mechanism of embryonic diapause has been adequately studied precludes any broad generalization from what has been observed. However, the biochemical features of embryonic diapause are quite similar to that of diapause in later stages of ontogeny.

For instance, diapausing embryos of the grasshopper *Melanoplus differentials* show not only a U-shaped respiration curve which is quite similar to that in the pupa, but also a reduced susceptibility to HCN and CO which again disappears with the onset of development Similar characters have been found by Wolsky (1941) in diapausing eggs of *Bombyx mori*. Diapause in the eggs of migratory locusts, as of *Locustana pardalina* , is of a markedly different character, and it probably resembles more to the type associated with AH-MH deficiency.

We are gratified by the work of Fukuda (1952a-c, 1953a-c) and Hasegawa (1964) and their colleagues for the first trustworthy information on the hormonal mechanism of embryonic diapause. These papers are, however, not concerned with diapause itself, but with the way it is determined in the mother organism. Since long time it has been accepted that the occurrence of diapause in the bivoltine races of the silkworm is determined by the environmental conditions under which the egg of the parent generation developed. But, the environmental conditions do not, as might perhaps be conjectured affect the rudiments of the gonads directly, but indirectly through the secretory functions of the nervous system which is yet undeveloped when determination occurs. The correct stimulus to the ovaries occurs not till the subsequent pupal period, as Umeya (1936) demonstrated for the first time in his ovary transplantation experiments. The complex mechanism whereby this transmission occurs was first discovered, however, by the systematic research of the two afore-mentioned authors. The folowing paragraph states the summary of their findings.

It can be demonstrated that the incidence of diapause in eggs is conditioned by the endocrine activity of special neurosecretory cells in the suboesophageal ganglion of the mother specimen. The extirpation of this ganglion at the beginning of the pupal instar causes eggs, which were determined for diapause through their mother, to develop without diapause. Similarly univoltine females can be caused to lay non-diapausing eggs, a thing which they never do under normal conditions. Conversely, if, up to a critical point in time, a female pupa which has been conditioned to lay non-diapausing eggs receives a suboesophageal ganglion from a female which would have laid diapausing eggs, it also will lay diapausing eggs. The same effect can be produced by the implantation of suboesophageal ganglia of either females or males of a univoltine race. That the relevant DH is nonspecific for species, genus or family was shown by implantations of the suboesophaeal ganglia from other Lepidoptera. The capacity to induce diapause was found not only in the ganglia of other species with an embryonic diapause such as *Antherea yamamayi, Dictyoploca japonica* and *Lymantria dspar*, but also in the ganglia of *Antherea pernyi*, a species with a pupal diapause. Ganglia from pupae of *Philosamia cynthia* and *Dendrolimus pini* did not have the similar.

Yet more co plex is the mechanism of transmission. complicated. A female conditioned to lay non-diapausing eggs can be made to deposit diapausing eggs by decerebration or imply by cutting the circumoesophageal connectives before a certain time. From these kind experiments it was affirmed that the endocrine activity of the suboesophageal ganglion, which conditions diapause in the eggs, is constricted by a nervous stimulus from the brain transmitted through the circumoesophageal connectives.

Hasegawa (1957) stated that he has succeeded in isolating an effective DH extract from *Bombyx mori* pupae which had been conditioned to lay diapausing eggs (univoltine race). The suboesophageal ganglia together with the brains of about 15-000 pupae were dissected and homogenized in 80 per cent methanol. After contrifugation, the supernatant fluid was evaporated under vacuum, diluted with water and washed with ether. The aqueous layer was then extracted with chloroform. The evaporated chloroform extract yielded a fluid which showed a high diapause inducing activity when it was injected into female pupae conditioned to lay non-diapausing eggs. An extract was obtained by the same method, from the heads of adult females. Isobe *et al.* (1973) recently succeeded in isolating two DH components from the heads of two million mated adult *Bombyx* males. They were separated from a complex lipid fraction by gel permeation chromatography, using Sephadex LH-20 and organic eluants. Both the components, A and B, seem to have a relatively low molecular weight (between 2000 and 4000) and they both seem to be complex lipids with a number of peptide bonds among several types of amino acids and various other components. The process of extracting the hormone was described (the active principle was found in the aqueous fraction after ether extraction and was then removed from this with chloroform). Its physicochemical and chromatographic properties and its acid, base, heat and light stability were determined. The isolation technique is illustrated in Table 1.

More research was done by Hasegawa (1963) and Yamashita and Hasegawa (1964a, b) who further examined the mode of action of the diapause hormone in the silkworm (*Bombyx mori*). The fate of the hormone after injection into a female pupa and its effect on the penetration of 3-hydroxy- kynurenine into the developing egg are discussed. The latter substance in the precursor of the serosa pigment (ommochrome) determining the colouration of the diapausing eggs; it is much less abundant in non-diapausing eggs. More experiments where done to know about, the effect of the diapause hormone on the glycogen content of the ovaries and the blood sugar level of the female pupae.

It has been confirmed that one of the actions of the hormone was to stimulate the penetration of blood sugars into the pupal ovaries and their deposition therein the form of glycogen, thus reducing their content in the haemolymph.

Further break through in the investigation of the mechanism of embryonic diapause development in silkworms was made by Chino (1961). It was shown that almost all the glycogen present in the eggs of *Bombyx mori* before the onset of diapause was rapidly decomposed by a complicated enzymatic reaction, the end products of which were the two polyhydric alcohols sorbitol and glycerol. A resynthesis of glycogen started at the moment when diapause ended. It was further shown that sorbitol is formed by dehydrogenases catalysing the following two reactions:

$$\text{glucose TPNH} \rightarrow \text{sorbitol} + \text{TPN} \qquad (5)$$

$$\text{glucose (or fructose)-6-phosphates} + \text{TPNH} \rightarrow \text{sorbitol-6-phosphates} + \text{TPN}^+ \qquad (6)$$

Glycerol is formed by enzymes catalysing the following reactions:

$$\text{dihydroxyacetone phosphate DPNH} \rightarrow \text{a-glycerol phosphate} + \text{DPN}^+ \qquad (5')$$

$$\text{glyceraldehyde (or dihydroxyacetone)} + \text{TPNH-glycerol} + \text{TPN}^+ \qquad (6')$$

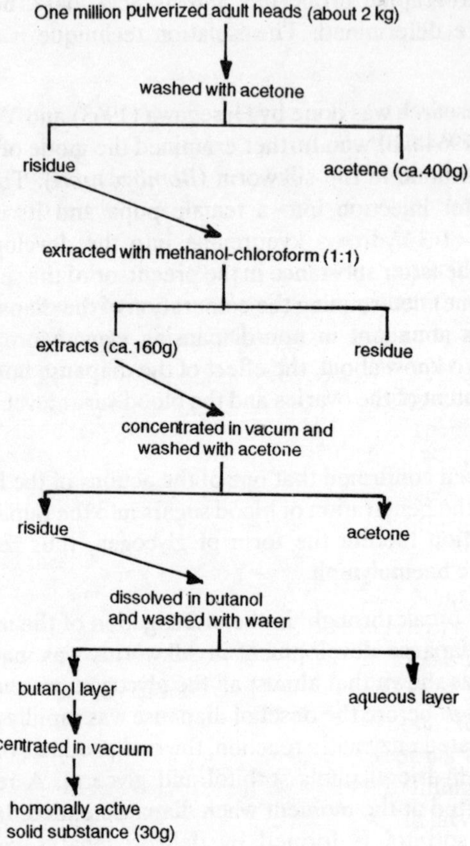

But these polyhydric alcohol dehydrogenases were not, the factors which induced diapause, for they were also present in developing eggs in which an increased amount of polyhydric alcohols never occur. The adaptive character of the accumulation of such substances which increase the resistance of diapausing eggs against freezing is clear.

It was discovered that if the crude diapause hormone extract was injected into female *Bombyx mori* pupae at the appropriate time, it not only induced diapause, but also assisted in the progress of glycogen accumulation in the eggs and distinctly stimulated trehalose activity in the haemo- lymph, resulting in a high glycogen level in the ovaries and its reduction in the fat body (Hasegawa and Yamashita, 1970). In another study, Kai and Hasegawa (1973) found that esterase A in Bombyx mori eggs was concerned with the diapause termination mechanism. When esterase A, separated by electrophoresis from non-diapause eggs or from eggs in which diapause had terminated, was added to an *in vitro* culture of *yolk* from diapausing eggs, the *yolk* cells were extremely affected (their cell membranes were dissolved, releasing the contents into the medium).

The synchronization of diapause in parasites with that of their hosts

Most of the researchers have confirmed that the endoparasitic larvae of entomophagous insects have a life-cycle closely

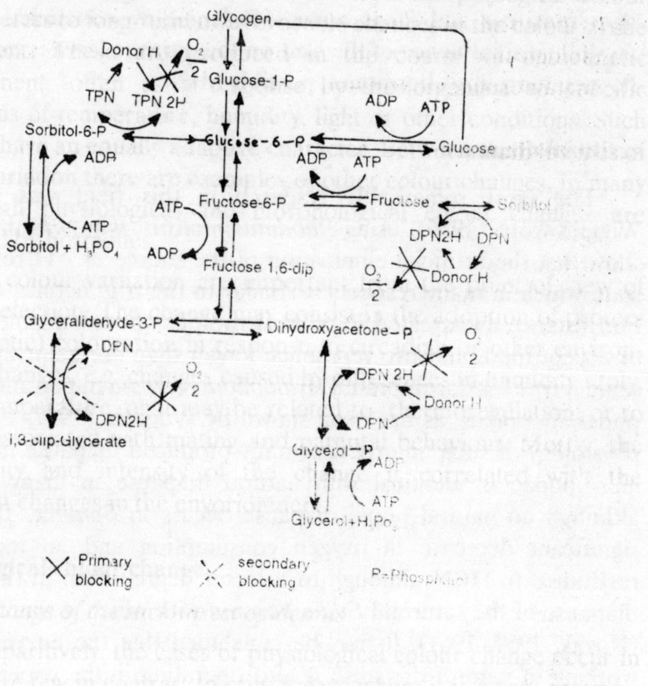

synchronized with that of the host. With small entomophagous parasites such as some Braconids and Tachinids, if the host larva is to survive and develop to an adult, such synchronization appears to be necessary. A special internal mechanism must be supposed in cases where both parasite and host undergo diapause. In this connection several authors have suggested an intervention by some of the metamorphosis hormones. This possibility has been studied experimentally by Schoonhoven (1962) in *Bupalus piniarius* L. (Geometridae) and its parasite *Eucarcelia rutilla* Vill. (Tachinidae). A series of experiments involving chilling, decerebration and parabiosis of diapausing host pupae has provided evidence that one of the two metamorphosis hormones associated with pupal diapause (AH and MH) is responsible for activating the parasite when the imaginal development of the host starts. Schoonhoven explained why this hormone must be MH. Apparently, the parasite is more easily affected by the hormaone than the host.

Maslennikova (1961) obtained similar conslusions while working on the development of the Chalcid, *Pteromalus puparum*, and its host *Pieris brassicae* and *Pieris rapae*. Maslennikova concluded that the physiological influence of the host on the parasite larva is exerted via the metamorphosis hormones, possibly MH.

Artificial diapause

The most remarkable research in this field was done by Wigglesworth (1936), using *Rhodnius prolixus*, who was the first to show that the artificial elimination of the source of AH results in a state which in many respects is similar to the true natural diapause. For instance decapitated *R. prolixus* nymphs, though unable to move or accept food, may live very much longer than normal nymphs. The same is true for decerebrated *Hyalophora cecropia* pupae and isolated pupal abdomens, as has been shown by Williams (1947, 1952). The physiology of such an experimentally produced diapause in *Bombyx mori* pupae is identical with natural diapause in many respects, although no natural pupal diapause occurs in *Bombyx*. There is a significant decrease in oxygen consumption and an increase in resistance to HCN although to a lesser degree than in the natural diapause of the saturniid *Eudia pavonia* with which it was compared (Novak, 1959, 1961). It may be concluded that the enzyme system working in natural diapause is also operating in the experimentally induced case, even though it may be much less developed. This suppports the theory that the origin of diapause was the result of a gradual evolution (see later), under the action of natural selection and concludes in an adaptation to adverse condition of environment.

7

The Metamorphosis Hormones

The insect hormones are those which are most effective and also well known by entomologists influence postembryonic development. At Till now, three of these so-called metamorphosis hormones are recognized: (A) the activation hormone manufactured by specific neurosecretory cells in the insect brain, which conditions the reactivation of the body after each moult and the production of the other two metamorphosis hormones; (B) the moult- ing hormone produced by the prothoracic glands or corresponding tissues, which conditions the moulting process and thus, indirectly, growth and morphogenesis; (C) the juvenile hormone produced by the corpora allata, which conditions the growth of the larval parts in all metabolous insects (Pterygota), the function of the ovarial follicles of adult females in most insects, and many other structures and functions of the insect body, which are unable to evolve in its absence. All three hormones are essential for the normal course of postembryonic development. The activation hormone is a typical neurohormone, the other two are true glandular hormones order, and sometimes even class unspecific. They are dependent on one another both in their production and function.

Terminology

Inspite of the fact the problem of insect hormones and metamorphosis has been intensively studied in recent years, or perhaps for this reason, no generally accepted terminology has been developed. Therefore it is better to define terminology used in the present work.

Metamorphosis

Metamorphosis implies that the single part of the postembryonic development which takes place in the absence of a morphogenetically active concentration of juvenile hormone, i.e. the last larval instar of Hemimetabola and both the last larval and the pupal instars of Holometabola. This is corroborated with most other physiological works but in disagreement with some morphologists. But it must not be denied that a part of the adult differentiation is

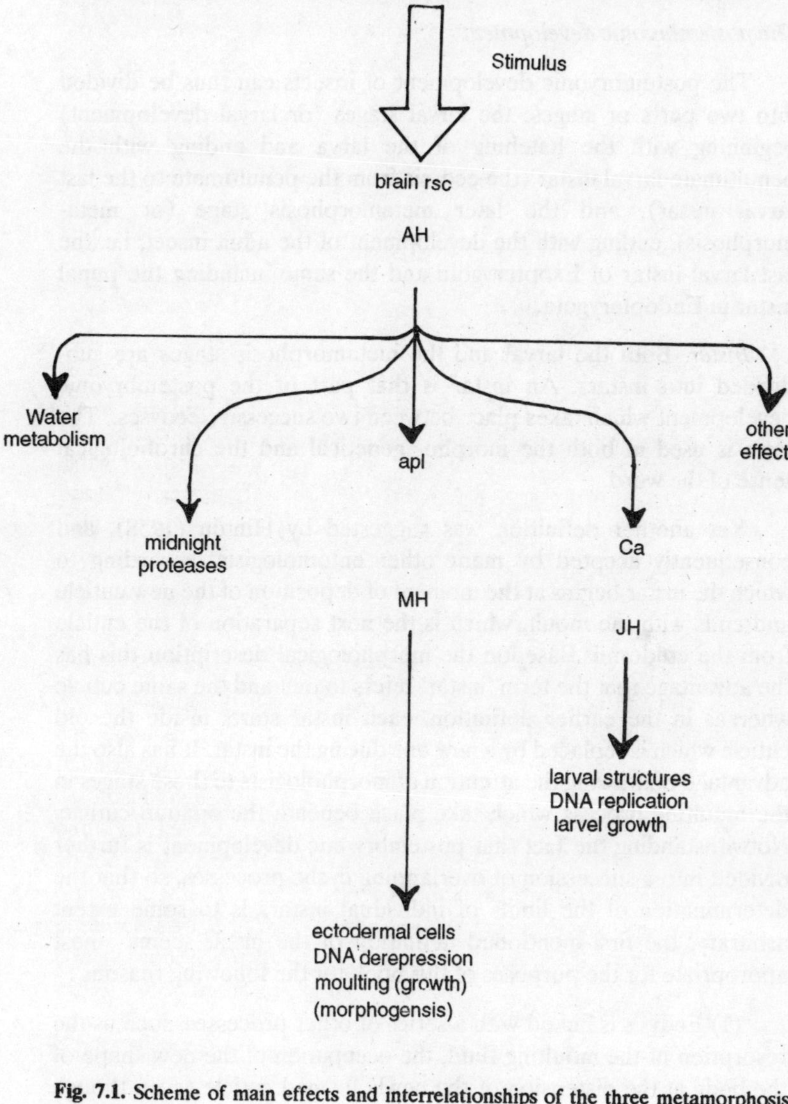

Fig. 7.1. Scheme of main effects and interrelationships of the three metamorphosis hormones..

recognized at the beginning of each larval instar, there is no more reason to call this metamorphosis than there is in the case of the similar successive postembryonic differentiation in ametabolous insects (Apterygota).

The postembryonic development

The postembryonic development of insects can thus be divided into two parts or stages: the larval stages (or larval development) beginning with the hatching of the larva and ending with the penultimate larval instar (the ecdysis from the penultimate to the last larval instar); and the later metamorphosis stage (or metamorphosis), ending with the development of the adult insect, i.e. the last larval instar of Exopterygota and the same including the pupal instar in Endopterygota.

Instar. Both the larval and the metamorphosis stages are subdivided into instars. An instar is that part of the postembryonic development which takes place between two successive ecdyses.. The term is used in both the morpho- genetical and the chronological sense of the word.

Yet another definition was suggested by Hinton (1958), and consequently accpted by many other entomologists, according to which the instar begins at the moment of deposition of the new cuticle and ends with the moult, which is the next separation of the cuticle from the epidermis.Basedon the morphological description this has the advantage that the term 'instar' refers to one and the same cuticle whereas in the earlier definition, each instar starts inside the old cuticle which is replaced by a new one during the instar. It has also the advantage of drawing the attention of morphologists to those stages in the moulting process which take place beneath the opaque cuticle. Notwithstanding the fact that postembryonic development is further divided into a succession of overlapping cyclic processes, so that the determination of the limits of individual instars is to some extent arbitrary, the first-mentioned definition of the instar seems most appropriate for the purposes of this book for the following reasons :

(1) Ecdysis is linked with a series of other processes, such as the resorption of the moulting fluid, the occupation of the new shape of the body at the distension of the newly formed cuticle (even though this was laid down some time previously), the deposition of the wax

and cement layers of the epicuticle, the tanning and hardening of the new cuticle, etc. It is therefore the most signigicant and visible section of each moulting cycle.

(2) Inspite of the fact that successive moults are connected with an uninterrup- ted succession of various physiological processes, most of these are, nevertheless, clearly arranged in cycles within each interecdysis, but not respiration ad hormone production, resulting in swelling of the during moults. By selecting the detachment of the cuticle as the start of an instar, all of these clearly logical processes are divided between two different instars.

(3) The structure of the body, even though determined by the newly deposited cuticle, particularly that of heavily sclerotized parts such as the head capsule, is also changed at each ecdysis but not at the moults.

(4) The detachment of the cuticle at apolysis can only be observed in histological sections (rather difficult to make because of the hard sclerotized cuticle). The moment of ecdysis is therefore as much preferable as the dividing point from a practical point of view.

Most of these reasons remain acceptable even in such cases as the last larval instar of higher Diptera. There, although the chief character- istic of ecdysis, the shedding of the exuvium, is eliminated, as is the tanning of the newly formed pupal cuticle, other distinguishing features, such as, for example, the removal of the moulting fluid and the formation of the epicuticular layers, remain; and these are enough to enable the instars to be distinguished. A similar conclusion has been reached by Wigglesworth (1973).

Pharate instar. If we do not accept Hinton's definition of the term 'instar', the term 'pharate instar' has a restricted meaning. It seems to be justified only in cases in which a new instar has been formed completely and the moulting fluid has been absorbed, as in the puparium of higher Diptera, for example. The determining factor is thus the extistence of air between the old and the new cuticle.

Development stage

Generally the term stage of development is applied for a section of ontogenetical development characterized by significant biochemical or morphological processes or by both, such as the protopod, poly- pod, oligopod, and postoligopod stages in embryogenesis, the larval stage (comprising the first to penultimate larval instar), the meta- morphosis stage (comprising the last larval and the pupal instars), the maturation stage, the stage of sexual maturity, etc.

Moulting process

A series of biochemical, physiological and morphological changes preceding each ecdysis constitutes the moulting process or *moulting*. This begins with the swelling of the epidermis followed by detachment of the old cuticle, usually referred to as the *moult* or *apolysis*. Then the moulting fluid is produced, in which the inner layers of the cuticle are gradually digested, and a new cuticle is laid down by the epidermal cell. The process ends in *ecdysis* which is preceded by the resorption of the moulting fluid and followed by the tanning and hardening of the exocuticle and the secretion of the wax and cement layers of the epicuticle. The laying down of the inner layers of the endocuticle by the epidermal cells persists after the ecdysis and usually does not stop until immediately before the start of the next moulting process. Thus ends the moulting process.

Few years back researchers began to use the terms 'larval-larval moult' in place of 'larval moulting', 'larval- pupal moult', instead of 'pupal moulting' and 'pupal-imaginal moult' instead of 'imaginal moulting'. Apart from the linguistic aspect, this seems completely unnecessary, in so far as we take into account that the above terms express the result of the relevant moulting process. In any case the meaning is usually amply evident from the cortex. However, the term 'emergence' for imaginal ecdysis is extirely adequate.

The term 'moult' does not mean 'moulting process' since it has been accepted as the English term for 'apolysis', i.e. only one of the initial phases of the moulting process.

Larval moulting-any moulting which results in larval form.

Pupal moulting-moulting to pupal form. when a second pupa is produced as a result of experimental JH treatment, we term it as 'second pupal moulting'.

Imaginal moulting or emergence–moulting which culminates in the adult form.

Growth

By growth is generally accepted as the increase in the amount of living matter of the body (organized protein molecules) unlike, e.g., the increase in weight by accepting food or ingesting water. The following lines piont the different types of growth may be distinguished in insects:

Isometric growth (harmonic or proportionate growth). The type of growth where there is an equal rate of increase in all parts of the body, i.e. there are no morphological changes. The shape of the body at the end of an isometric growth period is a homothetic figure of that at the beginning.

Anisometric growth (disharmonic or disproportionate growth). The type of growth where there is a different rate of increase in different parts of the body, with a resulting changes of shape. Two kinds of anisometric growth are distinguished in this book (cf. Novak, 1960, 1962). These are:

(i) *Allometric growth* (or anisometric growth in the restricted sense). IN this kind of growth parts of the body grow although at different rates, as for instance in postembryonic development in mammals.

(ii) *Gradient growth*. In this type of growth, only few parts of the body, the gradients, grow. Growth of the remaining parts is stopped and eventually they are wholly or partly removed either by histolysis or phagocytosis. This occurs, for example, in the embryo- genesis of most animals an in the metamorphosis of insects. Gradient growth is the significant feature in morphogenesis.

Differentiation

The important aspect of development, the sequence of changes which occurs in the body in the given period of ontogenesis, is irrespective of growth. The three following types are discerned (1) *chemical* (biochemical) differentiation, the sequence of changes in the chemical composition of the body irrespective of changes of weight, shape, function, etc., (2) *morphological* differentiation or morphogenesis, the sequence of changes in the shape of the body, (3) *functional* differentiation, the sequence of development of various functions of the body. These three types of classifications are all closely related in normal development. In ontogenesis, chemical differentiation usually precedes morphological differentiation, which in turn precedes functional differentiation. On the other hand in phylo- genesis, functional differentiation usually precedes chemical and morphological differentiation of the corresponding part of the body. (This follows from the action of natural selection and its role in evolution: it means that structures of the body can be favoured by selection and thus evolved and specialized for a given function only when this function already exists and is to some extent useful for survival to some extent.)

Larval structures. They are the parts of the body which can grow only the larval stage (see above), i. e. in the presence of a certain minimum concentration of juvenile hormone. They lose the ability for further growth during metamorphosis, and usually undergo partial or complete histolysis at that time (cf. Novak, 1951b, 1956). They include the greater part of the larval body in most pterygote insects.

Imaginal structures. Those parts of the body the growth of which is independent of the presence of juvenile hormone in the beginning of metamorphosis. Their development is accelerated during metamorphosis when the growth of the larval structures ceases. They include the greater part of the imaginal body.

Larvo-imaginal structures. Are fully differentiated and functional in the larval period, but they remain and continue to function in adults.

Larva

The term larva is used here in its broadest sense, as in Wigglesworth (1939) and Hinton (1958), but not Jeschikov (1941) and Snodgrass (1954), to include the entire postembryonic stage of development which precedes metamorphosis in all insects. The term 'nymph', as a special type of larva, is retained for the larval stage of *Hemimetabola* and 'larvula' is kept for the first larval instars of *Odonata* differing from the later ones. **'Eonymph', mesonymph'** and **'pronymph'** in the sense of Berlese (1913), and Slama (1960) refer to the pre-pupal stages of sawflies. Pre-pupa is the term for the latter portion of the last larval instar of *Holometabola*, which is distinguished by the advanced pupal moulting process with plenty of moulting fluid and retraction of the body.

Pupa

The term pupa is used for the last postembryonic instar (second meta- morphosis instar) in Holometabola, which is the period between the pupal and imaginal ecdyses. Internal morphological changes, which vary considerably in extent according to the insect, occur during this instar, whilst most of the external changes are occur in the first metamorphosis instar (last larval instar).

Hemimetabola and Holometabola

These terms are used as in the taxonomic sense of Imms (1946) and Obenberger (1952), but not Weber (1954), Snodgrass, etc. *Hemimetabola (= Heterometabola, = Exopterygota*) have only one metamorphosis instar, *Holometabola* (= Endopterygota) have two. The prefix 'hemi' does not refer to the degree of morphogenesis which, on average, is notmuch different in the two subclasses, but to the number of metamorphosis instars.

Abbreviations

Abbreviations are generally used for frequently recurring terms, for the purpose of saving space and also for greater clarity. The advantage of this was soon realized by chemists, who introduced a generally recognized and employed letter--symbol for every element (chemical compound). The same need is increasingly felt by specialists in insect endocrinology, but no rational abbreviation system which would be generally accepted has so far been created.

The system applied throughout this book, has the following rules Capital letters are employed for the hormones forming the main subject-matter and small letters for the often no less useful abbreviations of various endocrine organs. Differences in the origin of the same principle are expressed by adding a small letter or letters after the hormone symbol. The same primary symbol is always used for one principle (substance, effect). Consequently, instead of CAH, used by some authors for the corpus allatum hormone, and JHa for JH analogues, the common symbol JH is used for both, JH_{ea} denoting corpus allatum JH and JHa juvenile hormone analogues. JH is used alone where effects of the principle are discussed in general, irrespective of its origin. The abbreviations used by other authors are given in the list of synonyms at the beginning of each section dealing with the relevant hormones.

ca-corpora allata (corpus allatum).

cc-corpora cardiaca (corpus cardiacum).

nsc-neurosecretory cells.

ncc I, II, III-nervi corporum cardiacarum, first, second, and third pair.

agl-apolytic glands.

pgl-prothoracic glands.

rgl-ring gland.

vgl-ventral gland.

nho-neurohaemal organs.

mnho-metameric neurohaemal organs.

cns-central nervous system.

No abbreviations are used in this book for organs not associated with endocrine activity.

AH-activation hormone.

JH - juvenile hormone.

JH_{ca}-corpus allatum hormone.

JHa-juvenoids (juvenile hormone analogues).

MH-moulting hormone (ecdysone)

MH_{pg}-prothoracic gland MH.

MHd-ecdysoids (ecdysone derivatives).

GF-gradient factor (the distinction formerly made between the GF of insect metamorphosis and the general gradient factor gf has been abolished, because it is now assumed that they are identical).

BF-blastomo-factor.

To facilitate easy reading the terms are also given together with the abbreviations by which they are represented, wherever they appear for the first time in each chapter.

THE ACTIVATION HORMONE (AH)

(= 'moulting hormone', part. Wigglesworth, 1934; = 'Larvenhautungs-hormone' = 'Verpuppungshormon' = 'Imaginalhautungshormon', part. Piepho, 1938; = 'Growth and differentiation hormone', part. B. Scharrer, 1948; = thoracotropin', B. Scharrer, 1952; = 'adenotropes Gehirnhormone', Weber, 1954; = 'Activationsfactor', Wigglesworth, 1952; = 'brain hormone', Williams, 1947, Gersch *et al.*, 1973)

The initial statement regarding a hormonal function of the insect brain was made by Kopec (1917, 1922) whose experiments with *Lymantria dispar* caterpillars proved conclusively that the brain is necessary for pupation. This was the first proof for an internal secretion of any kind in insects. Tauber (= Taabor) (1925), working independently, concluded, from several years' work on blood transfusion and ligaturing in *Deilephila euphorbiae* caterpillars, that insect metamorphosis is controlled by endocrines. More indication for the presence of the humoral control of moulting in other species of Lepidoptera is contained in the papers by Koller (1938) and Buddenbrock (1950).

But the confirmation of the existence of a special hormone, produced by the brain which controlled moulting, was achieved by Wigglesworth (1936) from his classic experiments with decapitation, parabiosis and transplantation in the blood-sucking bug *Rhodnius prolixus*. This was the genuine beginning of insect endocrino- logy which has now evolved into a special, very extensvely studied, biological discipline. It was the same author who, in collaboration with Hanstrom (1943), first discovered the source of the activation hormone in the neurosecretory cells of the brain.

Another comtemporary entomologist Fraenkel (1935) demonstruted a similar hormonal control of pupation in *Calliphora* larvae by his well-known ligaturing experiments. Similar conclusions were reached by Bounhiol (1938) from his experiments with *Bombyx mori*; by Bodenstein (1938), on the basis of several years' experimental work on the transplantation of appendages from young to fully grown cater- pillars; by Kuhn and Piepho (1940) for *Galleria mellonella*, and by Hadorn (1939), Hardon and Neel and by Hadorn and B. Scharrer (1938) for *Drosophila*. Since then the number of papers on endocrine control of moulting has increased in number.

Fukuda (1944) was the first to prove that the AH does not affect the moulting process directly but does so by controlling the secretion of the prothoracic glands. These findings were confirmed and closely examined by Williams (1952) and his colleagues.

The suggestion of the possible function of the corpora cardiaca in neurosecretion was initially made by De Lerma (1933, 1934) and its connection with the neurosecretory cells of the brain was elucidated by Hanstrom (1942), by Pflugfelder (1937, 1938c, d) and, particularly, by B. Scharrer (1937, 1941) and B. and E. Scharrer (1937, 1944) who confirmed it by a detailed histological and experimental study, that the neurosecretory cells of the brain with their axons, together with the corpora cardiaca, form a single neurosecretory system corresponding with that of the neurosecretory cells of the hypothalamus with their axons and the neurohypophysis, in vertebrates. Since that time the neurosecretory function of the insect brain and corpora cardiaca, and its connection with moulting and metamorphosis, has been found whenever it has been seriously looked for, that is, in practically all the chief orders of metabolous insects (Pterygota). There is also histological evidence for its existence in Apterygota.

Number of resarch papers by Gersch and his school on the chromato- phorotropic and myotropic activity of insect brain extracts has raised the question as to whether some of the newly discovered (e.g. neurohormone C and neurohormone D of Gersch und Unger, 1957a; cf. 1960) could be identified with AH. This problem has been discussed compreensively by Raabe (1959).

The possibility of the neurohormone D_1 and AH being identical was suggested by Gersch (1962) on the basis of experimental evidence. Recent investigations by the same author and his colleagues showed that the activation hormone is different from neurohormone D, and that it is composed of two components, with different effects. These are the activation factor I (AH_1), which stimulates the synthesis of every kind of RNA, and the activation factor II (AH_{11}), which raises membrane potentials and thus has a positive effect on permeability (Gersch *et al.*, 1969; Gersch and Sturzebecher, 1968, 1972; Gersch *et al.*, 1973; Baumann and Gersch, 1973). Both effects seem to be quite general in character, so that they might account for most of the specific AH effects, e.g. stimulation of the secretory activity of various glands metabolism, water balance, diapause development.

Nevertheless, it is yet probable that AH_1 and AH_{11} could be two separate neurohormones produced by different nsc.

A number of other functions have been attributed to many known nsc, both of the brain and single ganglia of the ventral nerve cord; these will be discussed later.

B. and E. Scharrer (1944) came to thee conclusion that a close analogy exists between the neurosecretory system of three large groups of animals, Crustacea, Insects and Vertebrates, are in full agreement with such a possibility. They confirm the earlier views of the above mentioned researchers, Hanstrom (1939), and others. The relationship between AH production and secretory activity of the corpora allata, which regulates the activity of the ovarial follicle cells (studied in detail by Johansson, 1958), is in agreement both with the general nature of the AH effects and with the features of neurosecretory system common to the three groups of animals mentioned.

The neurosecretory cells of the brain (nsc)

Morphology

The neurosecretory cells of the brain, the source of the AH, are usually arranged in two groups, placed symmetrically on the upper surface of each hemisphere near the median furrow in the pars intercerebralis protocerebri. There are 4 to 15 or more main nsc in each group (4 in *Bombyx mori*, 6 in *Calliphora* larva, 7 in *Pyrrhocoris apterus*, 7 to 8 in various Hymenoptera, 15 in cockroaches and up to several hundreds in Orthoptera). They are often quite visible in a living brain dissected in Ringer because of their milk-white or slightly bluish colour. The transparent nucleus often appears as a dark spot in the middle of each cell. The cells are more obvious when examined under dark-ground illumination, appearing shining white against the dark background. The axons of these two groups of cells form a pair of nerves-the nervi corporum cardiacarum 1 (interni). These two nerves cross inside the brain in the mid-line and each enters the corpus cardiacum on the opposite side of the body. External to each of the two above- mentioned groups of nsc is, in most insects, a smaller group, usually of 2 to 4 cells, the axons of which form the nervi

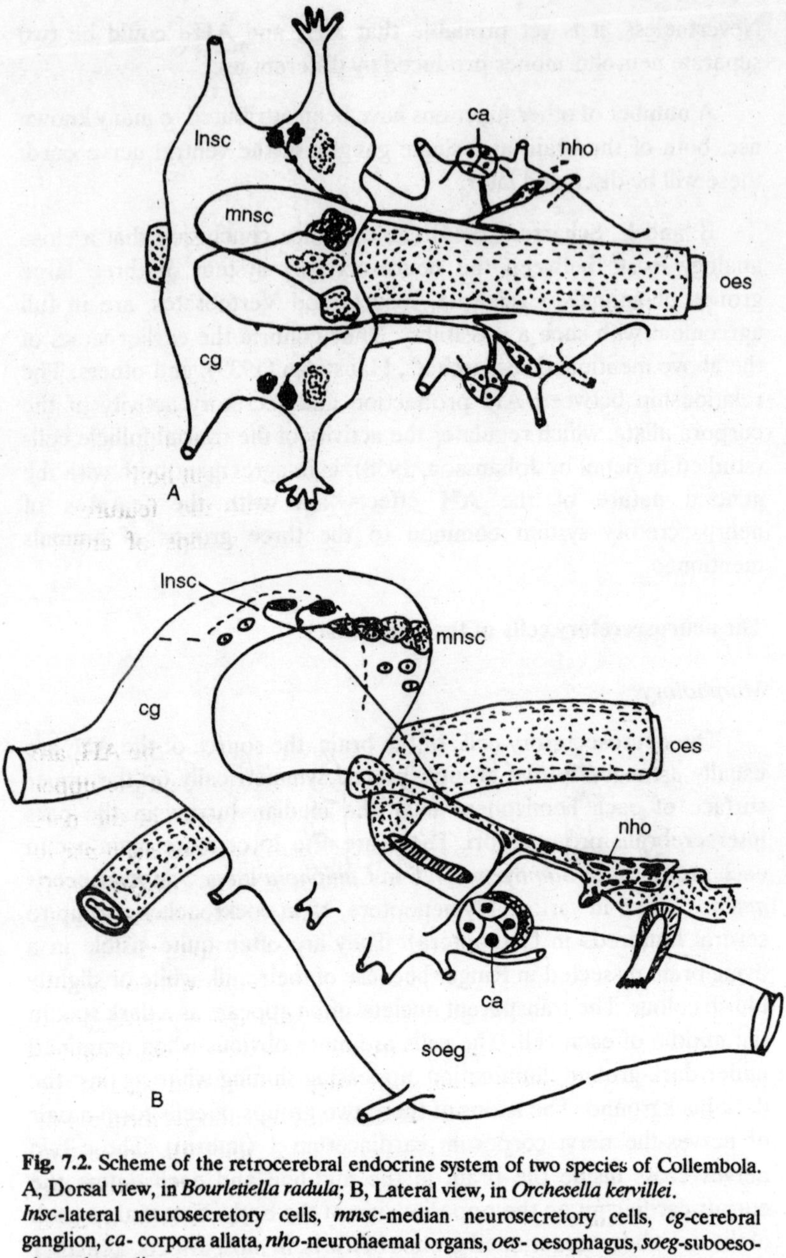

Fig. 7.2. Scheme of the retrocerebral endocrine system of two species of Collembola. A, Dorsal view, in *Bourletiella radula*; B, Lateral view, in *Orchesella kervillei*. *lnsc*-lateral neurosecretory cells, *mnsc*-median neurosecretory cells, *cg*-cerebral ganglion, *ca*- corpora allata, *nho*-neurohaemal organs, *oes*- oesophagus, *soeg*-suboesophageal ganglion.

corporum cardicarum II (externi), each nerve entering the corpus cardiacum on the similar portion of the body.

Many entomologist have explained the transportation of the material originating in the nsc of the brain into the corpora cardiaca via the nervi corporum cardiacarum. The neurosecretory material has been detected in histological preparations and under dark-ground illumination. Severing the nervi corporum cardiacarum led to an accumulation of neurosecretory material in the distal end of the portion connected to the brain and the disappearance of granules in the end on the other side of the cut. Resembling discovery was made by E. Thomsen (1954), using dark-ground illumination, when the cardiac- recurrent nerve* of adult *Calliphora* females was ligatured with a silk thread. The passing of part of the neurosecretory material from the corpus cardiacum via the nervus allatus into the corpus allatum, where it is xpected to stimulate the gland and increase its size, has recently been studied by Nayar (1958). E. Thomsen (1954) also observed that the neurosecretory granules pass via the nervus

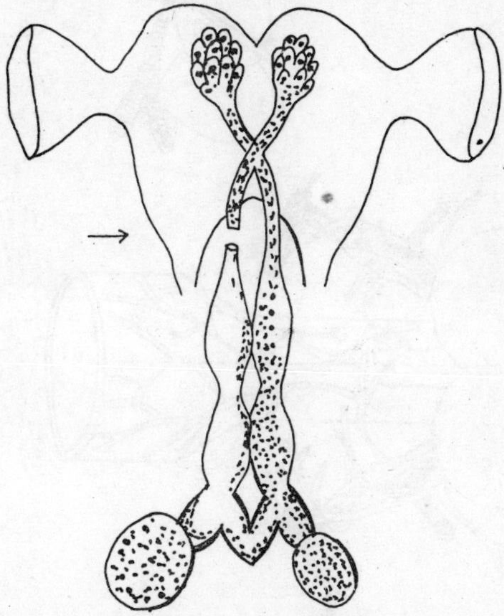

Fig. 7.3. The effect of breaking one of the two nervi corporum cardiacarum. Note the proximal accumulation of neurosecretory granules in the nerve stump, its disappearance from the corpus cardiacum and the enlargement of the corresponding corpus allatum.

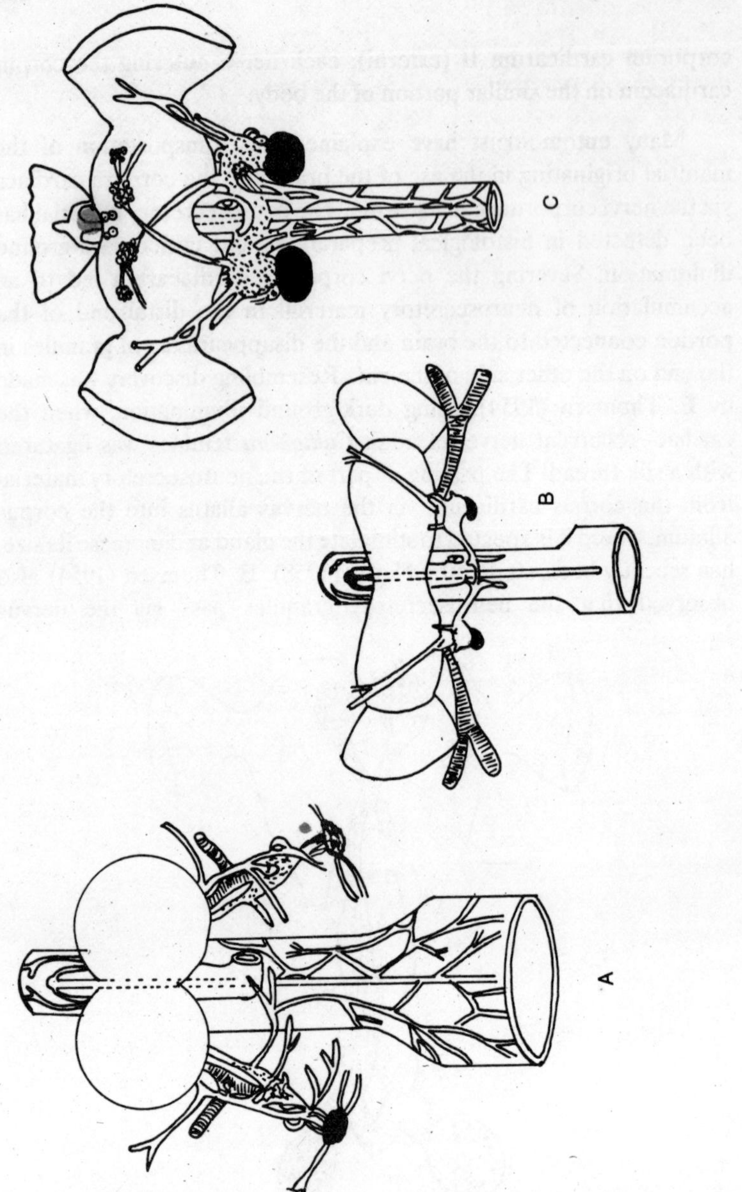

Fig. 7.4. Scheme of the brain and retrocerebral endocrine system in three developmental stages of Lemon Butterfly (*Papilio demoleus*). A, Last instar larva; B, Early pupa; C, Adult corpora cardiaca (dotted), corpora allata (black).

recurrence into the walls of the dorsal aorta, which is innervated by this nerve. From here the neurosecretory granules appear to more directly into the blood stream.

In many circumstances distinct cyclicity of the amount of neurosecretion was detected. For example Baehr (1969) found accumulation of neurosecretory granules in one of the three different types of A- nsc in about 12 min after a blood meal and their release at the of ecdysis, immediately after sucking blood and during egg laying.

Histology

The use of special staining techniques, such as Gomori (chrome haematoxylin-phloxin;) and paradlehyde fuschsin, renders the nsc exremely visible in histological preparations and the neuro- secretory material may readily be followed along their axons.

Using histochemical methods, a considerable amount of acid phosphaas was found in the cytoplasm of most of the nsc in the bug *Iphita limbata*, whilst it was nonexistent in the cytoplasm of the other nerve cells. On the other hand, the phosphates content of the nuclei of the neurosecretory cells was lower than that in the neighbouring nerve cells. Based the basis of his detailed histological studies on the brain of the lygaeid bug *Oncopeltus fasciatus* Dallas, Johansson (1958a, b) distinguished four different types of cells, the staining characteristics of which suggested a neurosecretory function. *A-cells*, the most conspicuous of all, stain dark purple-red or dark-blue with paraldehyde fuchsin and blue-black with Gomori chrome haematoxylin. They enclosed a large number of granules which fill most of the perikaryon. The cells are pear-shaped or sometimes irregular in shape, and occur immediately beneath the neurilemma. When the livecells were examined under dark-ground illumination, their shining bluish-white colour makes them quite visible. Their axons form the nervi corporum cardia- carum I. *B- cells* stain green or blue-green with paraldehyde fuchsin and red with Gomori. They are somewhat smaller than the A-cells and more irregular in shape. They are without granules and occur deeper in the mass of ganglion cells. *C-cells* stain purple-red with paraldehyde fuschsin and reddish with chrome haematoxylin-phloxin. They are found in the same layer as the B- cells, are irregular in shape and contain a flaky cytoplasm. *D-cells* have a fine granular cytoplasm which stains pale purple-red with

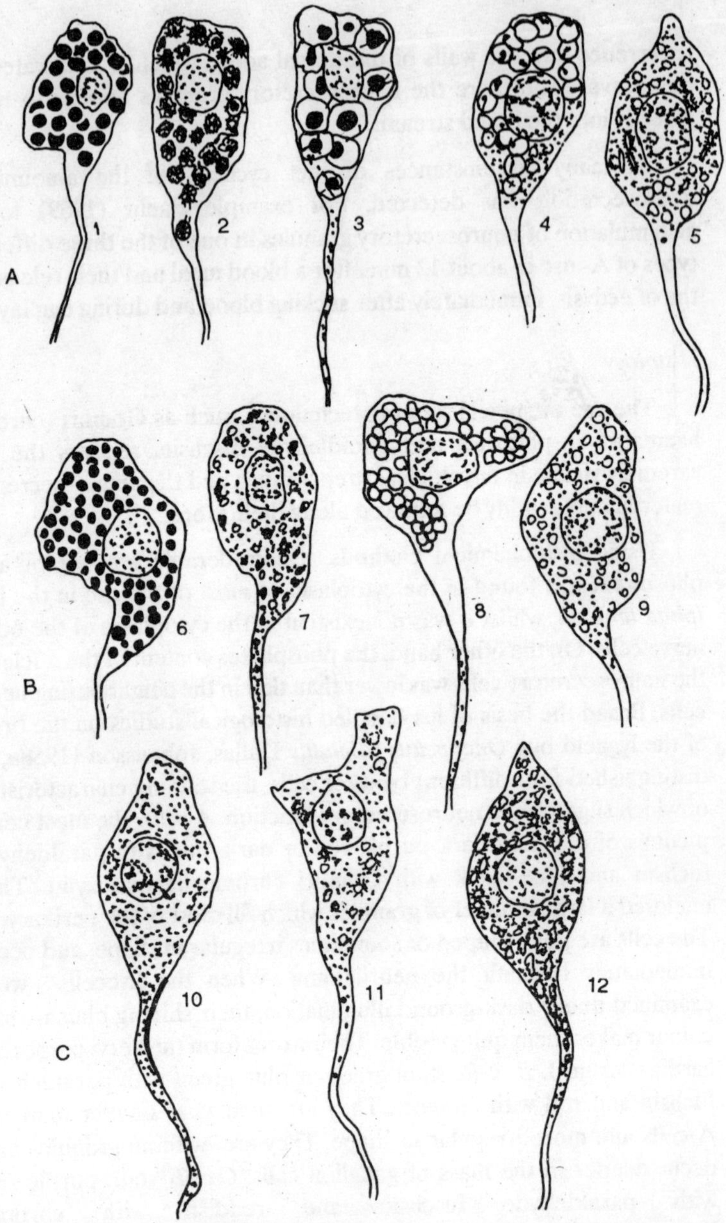

Fig. 7.5. Scheme of developmental changes in three types of neurosecretory cells of the pars intercerebralis in *Lampyris noctiluca*. A, nsc with large granules; B, nsc with medium sized granules; C, nsc with fine granules.

paraldehyde fuchsin and pale blue-black with chrome haematoxylinphloxin. They are to some extent larger than the A-cells. Removal of the A-cells resulted in reduced fertility, and delayed, but did nor prevent, egg laying. After extirpation the corpus allatum bwcame smaller.

Baehr, distinguished seven different groups, used different letter for the individual types, and so did Girardie and Girardie (1966), and Joly (1970) who followed Girardie. Changes in the histo- logical appearance of the neurosecretory system of the Cholorado Beetle *Leptinotarsa decemlineata* were similary detected by Schooneveld (1966-1972).

Ultrastructure

Because the ultrastructure of the pars intercerebralis was first described by B. Scharrer (1962) in the cockroach *Leucophaea moderae*, it has been well-considered by various other researcher e.g. Willey and Chapman (1962) in *Blaberus craniifer*, Normann (1965) and Bloch *et al*. (1966) in *Calliphora erythrocephala*, and Girardie (1967) in *Locusta migratoria*, etc. Diverse types of cells and neurosecretion were described.

Embryogenesis

THe lone information on the origin and development of the neurosecretory cells during embryogenesis are available in the papers by Jones (1953, 1956a, b). He discovered that in the embryos of *Locustana pardalina*, after diapause when the mitotic activity has occured again, several large cells can be viewed in those places in the brain where typical neurosecretory cells will later occur. These cells have a particular connection with the function of the ventral glands of the head and with moulting. They are probably the promary active sourcr of th hormone in the insect embryo.

Phylogenesis

There is plenty of research material on the source and evolution of the neurosecretory cells during phylogenesis. It has been that neurosecretory cells are found in all insects including Apterygota and also in other arthro- pods such as Crustacea Xiphosura Chilopoda Symphyla in Annelida and, phylogenetically 'lowest', in Turbellaria

Polyclada. Hanstrom (1940, 1953) presuppsed that neurosecretory cells originated with the so-called lateral frontal organs in the lower Crustacea, which lie in the epidermis quite separate from the brain. In the higher Crustacea they are very close to the brain and form the well-known pair of frontal organs. The neurosecretory cells of these are lined by their axons to the corresponding sinus gland at the side of the brain, where the secreted material is accumulated

In Apterygota, they form two group of cells on the surface of the protocerebrum which are enclosed by a connective tissue membrane. From all these groups a nerve arises which passes through the brain in a typical chiasma and reaches the corpus cardiacum of the opposite side. The neurosecretory cells and the granules in their axons give the same Gomori-positive reaction as the neurosecretory cells of Pterygota. The typical neurosecretory cells of Pterygota develop from these two groups by a process of inclusion into the brain mass during which the connective tissue membrane vanishes.

It is clear from the published resarched work that the activation hormone is phylogenetically the oldest of the three metamorphosis hormones. As regards the origin of the neurosecretory cells of the pars inter- cerebralis, Clark (1955) derives their function from the original epidermis cells from which they have evolved as nervous tissues and accepts their secretory function to be primary and their nervous function secondary. He indicates oints out that they are neurosecretory cells found in the most primitive, and phylogenetically oldest parts of the brain.

The corpora cardiaca (cc)

Morphology

It was Lyonet (1762). Who was the first mention of cc in the literature is without doubt that Lyonet (1762) in this work on the anatomy of *Cossus cossus*. They include of a pair of bodies situated immediately behind the brain, between the anterior end of the dorsal vein and the oesophagus, in front of the corpora allata with which they are linked by the nervi allati. They are sometimes fused medially and in some cases they are also fused with the corpora allata. In the ring-gland of Diptera Cyclorhapha, they form the lower median part of the ring, which is characterized by the small size of the cells, and fuse in the middle with the hypocerebral ganglion and at the sides with the R-cells. They are as mentioned earlier, innervated by two pairs of nerves from the brain, the nervi corporum cardiacarum I and II (interni and esterni) which are constitucted respectively of the axons of the median and lateral groups of neuro- secretory cells of the pars intercerebralis protocerebri.

The cc are connected medially with the hypocerebral ganglion by a nerve bridge and with the suboesophageal ganglion by the nervi allati. They send one or paired visceral nerves along the digestive tube to the visceral ganglia. The structure of the nervus allatus in *Gryllodes sigillatus* was described by Awasthy (1968) and in *Gryllus domesticus* by Belyaeva (1966), and by Theodorescu and Novak (1973). The cc is composed of two parts, a nervous part (originally, cc were supposed to be normal ganglia, the pharyngeal ganglia'), and a glandular part. They differ, at first glance, from ca and from all nervous ganglia by their bright, milky-white, opalescent colouration remembering that of the A-nsc.

Histology

Including the neurones (with scarce intrinsic neurosecretion) two other types of cell can be distinguished: (1) The chromophobic cells whose cytoplasm is to stained by the usual stain- ing methods. They consist the main connective tissue in cc in which there are, usually, scattered ceiks if the other type . (2) The chromo- philic cells which are deeply stained by various stains and furnished with characteristic pseudopodia-like processes. Cazal (1948) accepted them as modified neurones which had lost their nervous function and intensified their secretory function. In Plceoptera, Arvy and Gabe (1954) found small granules in the cytoplasm of the cc cells and in the nerve endings bringing the neurosecretion, which stained deeply with phloxin. Numerous ramifying fibers of the nervi corporum cardiacarum are present in the cc in addition to the alone-mentioned cells. Some of the granules produced by the neurosecretory cells of the brain are stored here, but the remainder reach the surface where they are washed off by the haemolymph. A small amount of neurosecretory material has been shown to pass through the cc into the ca, via the nervi allati in several species.

Besides to the neurosecretory granules, Nayar (1956, 1957b) found a secretion, which he did not accep to believe to belong to neurosecretory origin, in the chromophilic cells of *Iphita limbata*. De Lerma (1956) also advocates the existence of a special secretion of the cc in *Hydrous piceus*. It occurs in the form of homogeneous acidophillic inclusions which are somewhat abundant in the chromophilic cells of the cc. No intracellular Gomori-positive granules' were found within these two cell types of corpora cardiaca.

Arvy ad Gabe (1953a, b, 1954) also insisted on the occurrence of a true intracellular neurosecretion in the cc on the basis of their great knowledge of many groups of insects.

Johansson (1958) was able to differentiate three different types of cell in the cc of *Oncopeltus fasciatus*. Couple of these are large chromo- philic cells which occurs in different parts of the cc and differ in their staining characteristics with respect to paraldehyde fuchsin. Those of the frontal and lateral portions of the gland stain green whereas those of the hind and median portions take on a pale red colour. The third type are interstitial chromophobic cells. These are dispersed through-out cc but are most abundant medially. It is remarkable that Johansson did not find any accumulation of neurosecretory material in the cc of *Oncopeltus*, where the granules are stored only in the walls of the aorta dorsalis as in *Iphita limbata* changing from the condition in *Rhodnius prolixus* where they are stored in the cc. In *Hydrocyrius columbiae* granules are also stored in the walls of the aorta as they are in *Pyrrhocoris apterus*. Phylogenetically, it is apparently a secondary utilization of the tissue which capacites the neurosecretory material to pass into the haemolymph at a quicker rate.

Mayer and Pflugfelder (1958) discovered only two kinds of cells in cc of *Carausius morosus*: the osmiophilic cells containing granules of neurosecretory material and generally forming well-delimited groups in the corresponding part of the gland; and the osmiophobic cells containing a system of vacuoles, special lamellated granules and peculiar oblong mitochondria. In the nerve endings inside the gland, granules 500 to 2000 Å in size were found, which conforms to those known from the neurohypophysis of vertebrates and are of neurosecretory origin. M. Cazal (1971) studied the structure of the cc in *Locusta migratoria* in detail, while Divakar and Novak made an exact study of their morphology and histology in *Schistocerca gregaria*. Their histology during the penultimate and last larval instars, and the pupal instar in *Galleria mellonella* were indicated by Karachali and Novak (1973).

Ultrastructure

A much of the research to papers has been published on the ultrastructure of the cc in various insects species , the first being those by Pflugfelder (1958) in *Dixippus morosus* and by B. Scharrer (1963, 1968) in *Leucophaea maderae*. Other authors who examined the ultrastructure of the cc were Johnson (1966) in *Calliphora stygia*, King, Surrinder, *et al*. (1966) in larval *Drosophila melanogaster*, King, Aggarwal *et al*. (1966) in adult females of the wild type and a mutant of the same species, Normann (1965) in *Calliphora erythrocephala*, M. Cazal (1971), in *Locusta migratoria* and Cassier and Fain-Maurel (1971), in the same species, in is probably the most comprehensive and recent study on this subject. There are two methods in which neurohormones can be released into the circulating fluid, Cassier and Fain-Maurel regard the one which attributes the passage of intact granules by exopinocytosis as the more likely in *Locusta*. They observed no fragmentation of the granules and found four types of neurosecretory granules.

Function

Johansson (1958) assumed accepted thethe cc to has two functions: the of the products of diverse groups and types of neurosecretory cells, and in addition, the production of a separate hormone of their own. A similar suggestion was made by Cameron (1953) previously. He demonstrated that the cc produce a special active substance (orthodiphenol) which is quite different from neurosecretion and affects the heart-beat and peristaltic movements. Such a result was obtained by Altmann (1956), who examined the myo- tropic effects of corpora cardiaca extracts of the bug *Iphita limbata* on the gut peristalsis of another bug *Aspongopus janus* Fabr. and their chromatophorotropic effects on the red pigment cells of the decapod *Coridina laevis* Heller. The neurosecretory cells of the brain of the same species had no such effect. ohansson's (1958) experiments on the influence of the cc on reproduction failed to give postive results.

Recent discovery clarifies as to the intrinsic endocrine activity of the cc both in their glandular and their neurosecretory (intrinsic nsc) components, also as their neurohaemal function for the brain neurosecretory cells. These three functions are perrhaps all mediated by more than one active substance. In addition to AH_I and AH_{II}, cc were discovered; by many authors, to contain the following substance; polypeptides with myotropic and chromatophorotropic activity (e.g. the neurohormones C and D of Gersch and his school), hyperglycaemic and hypertrehalosaemic action produced by extracts of both neurohaemal and glandular components (so that at least two different substances are involved), substances affecting nervous function, substances having both a diuretic and an antidiuretic effect on water metabolism (most of them undoubtedly of a peptide nature, see M. Cazal, 1971) and a number of pharmacodynamic substances such as serotonin, orthodiphenol, etc. More information is required to identify all these substances and to consider their activity spectra.

Embryogenesis

The functioning of cc during the embryonic period was studied by Weismann (1926) and Pflugfelder (1937) in *Carausius morosus*; by Mellanby (1936) in *Rhodnius prolixu*; by Roonwal (1937) in *Locustana pardalina*; by Poulson (1937) in *Drosophila*, and by several otherresearchers. Their findings, cc, together with the hypo- cerebral ganglion, originate as unpaired dorsal evaginations of the oesophagus. Only later are they are separated and their cells differentiated into nervous and secretory ones. Initially, they are expected to have been normal visceral ganglia which secondarily attained a neurosecretory function and became innervated by the nervi corporum cardiacarum.

Phylogenesis

The cc are well formed in the Apterygota with the possible probable exception of *Collembola* (cf. Cazal, 1948). They are more developed in this group than the ca and have been described in Hapygidae and Lepismatidae. They are rather reduced in Campodeidae and are found in other Diplura and in Thysanura. It is probable that cc have developed during the phylogenesis of Apterygota, but earlier than ca. If Cazal's (1948) observation that the corpora cardiaca are absent in Collembola is confirmed, this will agree with the conclusion of B. Scharrer (1952) who comfirmed it from experimental evidence, that the cc merely store temporarily the secretion from the neurosecretory cells of the brain and are not necessary for the action of the activation hormone. Thus, for instance, transplantation of a brain with active neurosecretory cells may break diapause in *cecropia* pupae without simultaneous transplantation of the cc. There is as yet no specfic evidence for the presence of a special cc hormone as was stated by L'Helias (1956b).

The direct effects of activation hormone (AH)

AH is more complex in contrast with other hormones. It is difficcult to determine whether particular action of AH is related to that of the other brain neurohormones. This is often unasweable problem, so that the present classification is arbitrary and is based on the author's opinion, but supported by the data on AH_I and AH_{II} characteristics.

The activation of the prothoracic glands

The first rescarch papers by Kopec (1917) and Wigglesworth (1934, 1940) pointed out the necessity of the brain for moulting, pupation and metamorphosis. The discovery by Fukuda (1940) of the function of pgl in *Bombyx mori* demonstrated that the effect on moulting is an indirect one and depends upon the activation of these glands. This was confirmed by Williams (1952), who found that active pgl alone are able to induce moulting in isolated pupal abdomens of *cęcropia* while an active brain can only produce the same effect in the presence of (inactive) pgl. Wigglesworth (1952b) came to the same conclusion with *Rhodnius prolixus* and by several other authors with various insects.

Apparently some of the recent finding seem to contradict thisFor example, Johansson (1958), discovered that complete extirpation of all ten of the type A neurosecretory cells of the pars intercerebralis in *Oncopeltus fasciatus*, carried out not more than four hours after the last larval moult, did not prevent imaginal moulting, although this process started two or three days later than in the controls. Undoubtedly, this is because of the secretion of the A-cells, which is stored in the form of granules in the walls of the aorta, as mentioned above, and, also may be, in the A-nsc of the ventral nerve cord. According to Raabe (1959b), the way in which pgl are stimulated by

AH has not yet been explained. A possible explanation was suggested by Church (1955), who considers it might be directly concerned in the synthesis of ecdysone. The findings of E. Thomsen and I. Moller on the influence of AH on intestinal protease activity in *Calliphora erythrocephala,* as well as other effects, would however suggest an action of more ordinary nature, as would also its relationship with the antidiuretic hormone of the vertebrate hypothalamus.

The activation of the corpora allata

AH's effect on the function of ca is now adequately demonstrated. AH is necessary for the re-activation of ca in a manner similar to that of pgl. This was suggested in some of the earlier accounts and is quite evident from the work of Johansson (1958a, b) de Wilde (1958a-c) and his collabo-rators, Nayar (1957). Johansson (1958a) observed the effect of extirpation of the A-cells of the brain on the volume of ca and found that complete extirpation of these cells results in extraodinarily small ca. The same process was observed by E. Thomsen (1952) in *Calliphora* only a few A-cells are left, the ca volumes remain normal. The proper understanding of how the ca are activated by AH has proved difficult for two reasons: (1) The action of the ca seems to depend upon nervous stimuli; (2) the activation hormone reaches the ca not only through the haemolymph, but also directly from the brain, in the form of granules, via the nervi corporum cardiacarum, cc and nervi allati. Neither of these factors appears to play a primary role in the regulation of ca activity, except in the case of diapause (see below), but such has been attributed to both by different authors. The activation of ca produce the primary effect of the activation hormone on the ovaries.

The effect on ovarian development

Including its indirect effect on ovarian development through the activation of the ca, the existence of a direct effect of AH on the development of eggs has been hinted. A detailed examination of the endocrine mechanisms controlling ovarian development in adult *Calliphora erythrocephala* showed that removal of the neurosecretory cells of the brain can be extirely compensated for by the implantation of active nsc from another specimen but not by active ca. The effects of AH deficiency are different from those of allatectomy. Implantation of cc from an active specimen has a similar, though lesser, effect than the reimplantation of neurosecretory cells. The cc alone in this species. Thomsen concludes that the function of AH in the insect organism is quite complicated and that the nsc of the brain form an 'all-over controlling centre of the endocrine system'.

The removal of median nsc from the ovaries Indian Housefly *Musca nebulosa*, the ovaries failed to develop. The effects were not similar with those of allatectomy; instead of the hypertrophy of the fat body which follows allatectomy, the fat body was reduced in size after nsc removal compared with the accessory glands, and the accessory glands were devoid of secretion. This is probably absence of the general stimulant effect of AH. No yolk was deposited after median nsc removal in the flesh-fly *Sarcophaga bullata*, although yolk deposition started after their reimplantation, but not after the implantation of ca or of brain tissue minus the nsc. Removal of brain in female *Bombyx mori* pupae often caused the laying of diapause, instead of non-diapause eggs, which were distinguished by low carbohydrate and lipid concentrations.

The impact of the nsc of the pars intercerebralis on the *development of the ovaries* has been. Highnam (1962) has demonstrated that in *Schistocerca gregaria* they exert a positive control over oocyte development and that copulation as well as electrical stimulation, drastic wounding, or enforced activity may all bring about a release of material from the nsc of the pars inter-cerebralis and accelerate the development of the terminal oocytes in 14-day-old females reared without males. Hill (1962) found that in the same species a there is positive control of *haemolymph protein concentration* by neuro-secretion during ovarian development. The presence of mature males to accelerates the release of material from the female neurosecretory system, resulting in rapid development of terminal oocytes and an increase in the number of eggs. A positive effect of the pars intercerebralis on the development of the ovaries, as well as on postembryonic development, on the *function of the ventral glands* and ca, and on *metabolism, water balance*, and *chromatic adaptation* was described by Girardie (1964) in *Locusta migratoria*. Mordue (1965a-c) too, obtained similar result with adult females of *Tenebrio molitor*.

The effect on the accessory glands

The impact of cerebral neurosecretion on the male accessory glands, initially discovered by Wigglesworth (1936), and since then confirmed by various researcher working on other species of insects is, as in the case of the ovaries, primarily an indirect one through activation of the ca. No direct effects have obeen till now observed, though they may be assumed.

The control of diapause

As indicated by E. Scharrer (1952), the author of the concept of neurosecretion, the specific role of neurohormones in the animal body depends on a long-term co-ordination between the nervous system and the ininternal secretion. A classic example of this interdependence is the direct influence of AH on both larval (pupal) and imaginal diapause. In each case it is a nervous stimulus induced by external conditions which induces a temporary interruption of AH production which in turn results in the inhibition of the functions concerned.

As demonstrated by Williams (1948) and several other entomogists a stage corresponding to natural diapause can be caused by the removal of the brain, the source of AH. Similarly, a natural diapause can be broken at any time by implanting an active brain or other source of AH (corpora cardiaca or corpora allata; see below). It seems that this dependence is the result of a phylogenetic utilization and improvement by diapausing insects of a tendency present in all insects. This results in the control of AH production in response to a variety of unfavourable environmental conditions and it is highly adjustable.

Thus the process of diapause generally occurs in the winter period in temperate regions, whereas it is common in the hot dry season in the tropics and subtropics. The same is true of the dependence of AH production on the nervous impulse produced by the distension of the abdomen in blood-sucking insects as described by Wigglesworth (1936, 1948) and by Detinova (1954). The condition in blowfly larvae first observed by Cousin (1933) may be asssumed as an transitional stage in the developing process diapause..

The effect on morphogenesis

In the experiments with the cricket *Pteronemobius haideni*, Sellier (1956) found that the implantation of an active brain into the larva (VIIIth instar) at the start of the diapause not only in suppression of diapause, but also in a lengthening of the wings when they bwcome adults. But there were no changes when the implantation was effected at a more advanced stage in diapause. The implantation of brains from other species of the genus *Gryllus* gave similar results. However, this seems to be much more an indirect effect depending on the time of activation of the ca.

The entire problem of both direct and indirect effects has been discussed by Gilbert (1964) and Buckmann (1969), but many questions are yet unanswered.

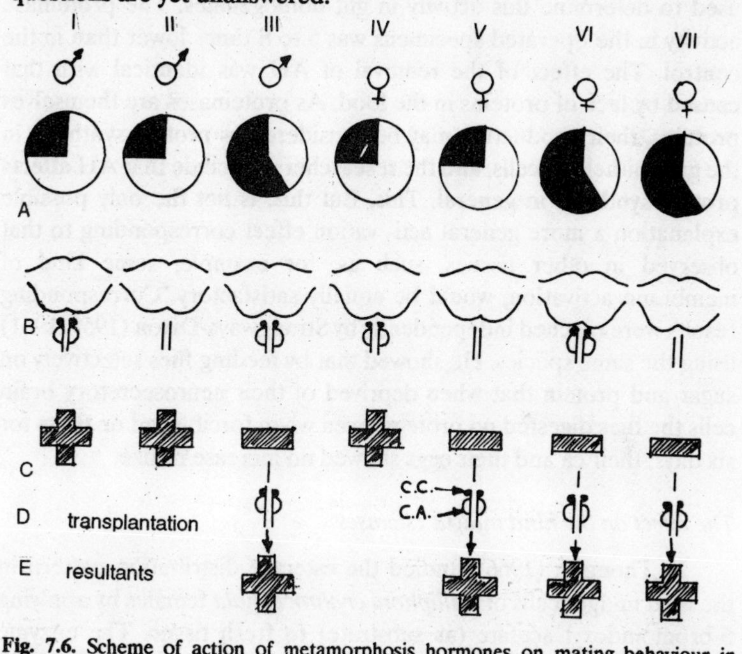

Fig. 7.6. Scheme of action of metamorphosis hormones on mating behaviour in *Pyrrhocoris apterus* adults. A, Length of daily photoperiod (black section = darkness); B, endocrine organs (*in situ or removed*); C, Sexual activity: + clear and significant, -weak or absent; D, + Gland implantation; E, Sexual activity after implantation. The figure shows that a short photoperiod may suppress sexual activity in males only, but that even in this case it reappears after cc and ca implantation. In females, both a long photoperiod and active cc and ca are necessary for sexual activity, but the implantation of active glands can restore activity in both cases.

The effect on the fat body

An observation made by E. Thomsen (1952) proves that the fat bodies of *Calliphora* females, whose AH source had been removed, had a higher glycogen content though the quantity of fat was reduced. The glycogen content corresponded with that following allatectomy but the reduction in the quantity of fat was greater.

The effect on intestinal proteinase activity

The two entomologist known as, E. Thomsen and I. Moller (1959a, b) studied the effect of extirpation of the neurosecretory cells of the pars intercerebralis on intestinal proteinase activity in *Calliphora* females. The colorimetric method of Day and Powning was used to determine this activity in gut homogenates. The proteinase activity in the operated specimens was 5 to 8 times lower than in the control. The effect of the removal of AH was identical with that caused by lack of proteins in the food. As proteinases are themselves proteins, their production may be considered as protein synthesis in the gut epithelium cells, and the researchers conclude that AH affects protein synthesis in general. This, But this, is not the only possible explanation-a more general acti- vation effect corresponding to that observed in other tissues, such as, for example, some kind of membrane activation, would be equally satisfactory. Corresponding results were reached independently by Strangways-Dixon (1959, 1961) using the same species. He showed that by feeding flies selectively on sugar and protein that when deprived of their neurosecretory brain cells the flies digested no proteins even when forcibly fed on them for six days; their ca and their eggs showed no increase in size.

The effect on the hind midgut esterases

E. Thomsen (1966) studied the esterase distribution pattern in the hind midgut cells of *Calliphora erythrocephala* females by applying 5-bromoindoxyl acetate (as substrate) to fresh tissue. The enzyme appeared in the form of granules, filaments and 'caps' similar to those described by Wigglesworth (1958) in *Rhodnius prolixus*. Generally in meatfed flies, the pattern of enzymes in contact with fat droplets followed a cycle correlated with ovarian development. In flies fed on a protein-free diet, the esterase level was low. The same finding was made in flies deprived of their medial nsc, even when they were fed on meat. It was comfirmed that the medial nsc control, or influence, esterase production in the cells.

The effect on water balance

Altmann (1956a, b) experimented on studied the effect of extracts from different endocrine glands on the intake, retention, and excretion of water by the honey-bee. Injections of ca extracts increased water consumption and decreased the viscosity of the haemolymph; cc extracts increased haemolymph viscosity and decreased water consumption. Excretion was increased by ca extracts and decreased by extracts of cc. Injections of adrenaline produced similar effects to corpora cardiaca extracts. The effects of the corpus allatum extracts could be due to juvenile hormone or to neurosecretory material reaching the gland via the nervus allatus; the latter is more probable.

In *Iphita limbata*, when distilled water was provided for drinking for some time an increased amount of neurosecretory material was found in the walls of the aorta, and nsc were practically free of granules. While in the insects which were forcibly fed on salt water for the same length of time, the release of neurosecretory material through the walls of the aorta almost stopped and the neurosecretory cells were filled with granules (Nayar, 1957). There were similar results were obtained with specimens where the wax layer of the epicuticle was removed by washing with benzene or chloroform. In insects kept in an atmosphere dried with calcium chloride, neurosecretory material accumulated in the nsc; in those kept in a more humid atmosphere, the material was released and enlarged quantity of granules was found in the walls of the aorta. Same result were acquired by Gutmann and Novak* in *Pyrrhocoris apterus*. It may be concluded that nsc enable the animal to maintain the water balance within certain limits. The comparison with the antidiuretic hormone of the vertebrate hypothalamus is striking.

Raabe (1959a) reached to the same conclusion with *Carausius morosus* where the cc had the same effect as the neurosecretory cells of the brain. This agrees with the findings of Stutinski (1952a, b). He injected rats with extracts of the pars intercerebralis protocerebri and cc of the cockroach *Blabera fusca* and found that the urine was reduced in quantity in the same way as after the administration of pitressin. The opposite result was obtained by Nunez (1956) using the beetle *Anisotarsus cupripennis*. There, the extracts of brain nsc and cc appeared to produce diuretic effects. These experiments, are yet to be analysed under comparable conditions.

Control of phase differentiation

Conforming with Girardie (1966) and Girardie and Girardie (1966), Joly (1970) assumed that phase differentiation in *Locusta migratoria* was controlled by special nsc in the pars intercerebralis. Three types of cells were recognized in the pars intercerebralis: A-cells, which are small and contain a large amount of Gomori-positive neurosecretion, B-cells, which are similar, but have highly phloxinophilic cytoplasm and no neurosecretory granules (A- and B-cells are regarded as being possibly two different stages of the same type of cell) and C- cells, which are much larger and contain only a few neurosecretory granules. But, no comparison with various types of cells in other insects, similarly experimented by other authors, is available. Destruction of the C-cells by electrocoagulation had the same effect as ca extirpation. If performed at the beginning of the IVth (penultimate) larval instar, the result would be a well-developed adult. Destruction of the A- and B- cells had the reverse effect, i.e. it acted like ca implantation. If carried out at the beginning of the last larval instar, a green adultoid developed. The implantation of extra A- and B- cells acted like allatectomy if carried out at the beginning of the penultimate instar. The researchers accepted that the C-cells activate the ca under normal conditions, while the A- and B-cells inhibit them. There is, however, a simpler explanation, consistent with findings in other insects, which has not been sufficiently ruled out, i.e. that it is injury to the C-cells which inhibits the ca in the first case. In the second case, removal of the A- and B-cells would prolong the interecdysial period of the last larval instar, so that the individual's own JH would, unlike in normal development, remain active; while the implantation of extra cells in the IVthe instar would shorten the intermoult period, with the result that the ca would not attain an active concentration in the penultimate instar. Joly's inability to find any difference between ca ultrastructure in normal specimens and those with destroyed A- and B-cells supports the second theory

Other effects

Many Confirmed effects of AH, as well as the possibility that they may be due to a common and very general action, leaves no doubt that further research will reveal a number of other effects. It also remains to be determined whether some of the observed effects of brain and cc extracts, such as the various chromatophorotropic and myotropic actions, are due to AH or to other neurosecretory products. There is little experimental evidence for any definite information on this connection.

The indirect effects of AH

Probably a hormone may act both directly and indirectly on the same tissue, e.g. it may stimulate RNA production by the tissue directly and at the same time induce the production of another hormone stimulating tissue growth or function. Another aspect to be considered is that the direct or indirect effects may be of different degrees. For instance, a hormone may act on the target tissue directly or may induce another hormone to do so, or the other hormone may cause a metabolic change in which only the tissue is affected.

The control of the moulting process must be considered as an indirect effect of the activation hormone. Moulting is restricted by removal of the AH source and re-evoked by its reimplantation. For this reason, the brain hormone was originally described as the moulting hormone by Wigglesworth (1934). It was called a 'growth and differentiation hormone' (*partim*) by B. Scharrer (1948) and Williams (1947, 1952a). The ca hormone was also initially interpreted by researchers as a moulting hormone. Confimating of an indirect as opposed to a direct effect is provided by the fact that the effect can be produced by another hormone but not by the given hormone alone. While a mere finding that removing another organ inhibits the effect is in no way proof of an indirect effect. In such a case there is always the possibility that the effect of the hormone was prevented by breaking another link in the chain between the hormone and the part of the body responding. With the metamorphosis hormones it is particularly important to differentiate between the direct and indirect actions of a given hormone. The interaction of their effects is very complicated, as in metamorphosis and moulting. View, a revision of many of the false on the metamorphosis hormones and their mode of action is required.

The control of AH production

The functional cycles of the neurosecretory cells. The existence of particular 'critical' periods in each instar as far as the hormonal activity is concerned was demonstrated experimentally shortly after the discovery of the first metamorphosis hormones. During these periods, the presence of the source of the given hormone is necessary for normal development whereas after wards it is no longer necessary. This shows that there is a certain cyclicity in the function of the AH, which has been confirmed by subsequent histological research on the neurosecretory system. The course of neurosecretion and the function of cc during postembryonic development has been studied in detail by Herlant-Meewis and Paquet (1956) in *Dixippus morosus*. They discovered a decrease in the amount of neurosecretory granules in the pars intercerebralis and cc at the time of each moult. Neurosecretory activity reached its maximum after about one-third of the next intermoult period had elapsed. Even at this time also, the largest amount of neurosecretory material was seen to pass into the corpora cardiaca. This period of maximum secretory activity agrees well with what has been observed regarding the critical period for AH release in other species. The quantity of neurosecretion in cc decreases makedly about half way through the intermoult period, i.e. when the critical period for the juvenile hormone has been reached. Another increase in neuro- secretory material was observed in the last quarter of the intermoult period. Cyclical changes in the amount and character of the neurosecretory material were observed in Phasmids and other insects by a number of other authors while a report from Fuller (1959) discrbes recurrent cyclical changes in the amount of secretion in the individual neurosecretory cells of the pars intercerebralis of *Periplaneta americana* without any noticeable cyclicity in the neurosecretory system as a whole.

There have been sugestions about the factors which govern the changes observed in neurosecretory activity. One of the first answers was provided by Wigglesworth (1936a, 1940a) in his classic experiments with the blood-sucking bug *Rhodnius prolixus*. He demostrated that production of the brain hormone and thus also the start of the moulting process, and each the interrelated developmental processes, is induced by a nervous stimulus produced by distension of the abdomen due to ingested blood. This stimulus depends on the quantity, not the quality, of the fluid intake, since the same effect is caused by a corresponding volume of pure water. Several small blood meals do not, however, produce any effect, even though their sum exceeds the necessary minimum quantity. Severing the ventral nerve cord anywhere between the brain and the abdomen prevents the passage of this stimulus. Therefore it can be accepted as a typical-case of the transformation of a nervous stimulus into an endocrine impulse effected by the nsc of the pars intercerebralis protocerebri, as claimed by E.Scharrer (1952). A very similar neurosecretory effect was found by Detinova (1954) in *Anopheles*. On the other hand, no such connection was observed by Novak (1951b) in the plant-feeding bug *Oncopeltus fasciatus* and many other insects where poorly fed specimens can undergo a normal moulting process. It is also recognized that meal-worms undergo several extra moults when left without food, and starving clothes-moth larvae can moult as many as 40 times. The above-mentioned dependence of AH production on a nervous stimulus from the distended abdomen is probably a secondary phylogenetic adaptation in blood-sucking insects. It enables them to survive for long periods in the absence of a suitable host in a state of reduced metabolism corresponding to that of a true diapause. In most other insects, AH production is automatically renewed at the beginning of each instar with the passing of the first digested food into the haemolymph. The phylogenetic character of this dependence on the stimulus from the abdomen was shown by Larsen and Bodenstein (1959) in the mosquitoes *Culex pipiens* and *Aedes aegypti*. Whereas *A. aegypti* and normal *C. pipiens* show the usual dependence of AH production (and through this juvenile hormone production which regulates egg development and oviposition) on the distension of the abdomen by blood. AH production by the autogenic form, *Culex molestus*, is free of this stimulus.

An exhaustive study of the function of the neurosecretory cells in insects which feed continuously throughout their growth has been made by Clarke and Langley (1962) using *Locusta migratoria* L. No changes were discovered in the amount of neurosecretory material in the median nsc or in their axons (nervi corporum cardiacarum), or in cc during postembryonic development or at different times during the intermoult period. The neurosecretory material is apparently produced continuously throughout growth in this species. As the AH is presumably not used during the ecdyses, the authors assume it accumulates in the haemolymph at these times. The increased concentration of the hormone in the haemolymph at these periods would then reactive agl.

A very remarkable association between the frontal ganglion and the production of neurosecretion was also found. Its extirpation resulted in the complete inhibition of further growth and moulting. The same effect was attained by cutting the frontal connectives, whereas cutting the recurrent nerve or the ventral nerve cord in front of the first abdominal ganglion had no effect. Histological examination of neuroendocrine system five days after the operation revealed a marked accumulation of neurosecretory granules in cc, the nervi corporum cardiacarum I but scarcely any in the nsc. No neurosecretory cells were discoverd in the frontal ganglion. The authors supppose that the frontal ganglion plays a part in transmitting the nervous impulses from stretch receptors in the oesophagus to the nsc of the pars intercerebralis which would thus connect the release of the hormone with the intake of food.

A momentary inhibition of AH production caused by various external impulses, the mechanisms of action of which are not yet entirely apprehencled, is the direct internal cause of all types of insect diapause except the early embryonic one. In addition to photoperiodism, temperature and the above-mentioned distension of the abdomen in blood-sucking insects, several other factors controlling AH production have recently been described. Therefore the interrelationship between juvenile hormone production and the carrying of oothecae by female cockroaches, incorrectly interpreted as a direct effect on corpus allatum secretion, is undoubtedly governed by neurosecretion. Similarly, neurosecretion, or AH production, appears to be the chief mechanism influenced by the quality the food as shown by E. Thomsen (1959) and Strangways-Dixon (1959); and the effect of feeding on protein synthesis is also controled in this manner.

Thomsen and Lea (1968) examined cyclic changes in the medial nsc of *Calliphora erythrocephala* under various conditions. The nuclei and nucleoli displayed cyclic changes in volume and in the amount of neurosecretion. Neurosecretory activity rose on the first day after emergence, with acceptance of a sugar diet; it then fell again but rose once more after the fourth day in connection with the beginning of meat eating and fell again after egg laying. Allatectomy reduced nuclear and nucleolar volume ,but the implantion of active ca renewed neurosecretory activity in such specimens. The authors concluded that the activity of the neurosecretory cells was regulated by ca, but the effect is undoubtedly an indirect one (possibly by transmission to some of the nerve endings connected with the medial nsc).

Chemical characteristics of the AH

The researcher also who was first to identify the chemical nature of the brain neurohormone was L'Helias (1955a, 1956a).* Her pterinederivative theory suggested the existence of a close relationship between AH and JH on a pterine basis. A few years later, however, Gersch and Unger (1962) showed that pterines had nothing in common with either of these hormones and suggested, on the basis of earlier papers that neurohormones were possibly of a peptide nature. This accords with phylogenetic considerations on their relationship to vertebrate neurosecretory material, the polypeptide character of which has now been fully accepted.

The discoveries by B. Scharrer (1952) and others appear to speak equally against the hypothesis of L'Helias (1956) on the interaction of the brain secretion with those of cc or ca (1) extirpation of the cc has no qualitative effect either on moulting or metamorphosis; (2) the effects of brain extirpation may be at least partly compensated by the implantation of active cc (3) the careful histological studies on the neurosecretory activity of the brain and cc in some Hymenopytera and Diptera carried out by M. Thomsen (1954a, b) give support to the conclusion that the neurosecretory material passes through cc directly into the haemolymph.

Few rearchers ignored the findings of Gersch's school or phylogenetic relationships into account, however, and expressed entirely different views on the chemical character of AH and other neurohormones. For example, isolation of the active ingredient of the AH has been reported by Kobayashi and Kirimura (1958). They used 8500 silkworm (*Bombyx mori*) pupae preserved in methanol 24 hours after pupation and centrifuged three times after homogenization 200 ml of the methanol solution obtained were concentrrated to 30ml and extracted with 145 ml ethyl ether. On evaporation, the ether solution yielded about 2 mg of an oily yellowish brown material; the evaporation temperature did not exceed 38°C. A injection of this substance into decerebrated permanent pupae, in which no ecdysone had previously been found, caused these to moult to adults 16 to 20 days later. It has been claimed by Kirimura *et al.* (1962), that the active principle of these extracts is cholesterol and its identity with AH was taken for grantend. This view has, not much assistance from other facts known about AH. Nevertheless, when correlated with the recent finding of Karlson and Hoffmeister (1963) that cholesterol is the precursor of ecdysone, it becomes of remarkably.

Carlisle and Ellis (1963) injected IVth instar nymphs of *Locusta migratoria migratorioides* with cholesterol dissolved in ether. The purified preparation and one of two samples of commercially available crystalline cholesterol were ineffective whereas the other commercial sample, hastened moulting by about 18 hours (PLV.01). The authors found' that pure cholesterol as such has no prothoracotropic effect, which they assume to be derived from impurities present in the sample concerned. They conjectured that the active ingredient is a steroid related to cholesterol, and that the natural brain hormone is similarly a steriod of this group, or possibly a triterpenoid of related configuration.

Even Gillbert (1964) expressed his opinion in favour of the steroid character of the brain hormone in his review of the question. He also agreed with the earlier authors in assuming 'a number of neurosecretory substances produced by the brain, the prothoracotropic effect being produced by one of them and the other effects of the brain neurosecretion by others'.

Another researchertook a different stance Novak agreed with Gersch (1962, cf. 1964) in his assumption of the polypeptide nature of AH and claimed that most of the known effects of the brain secretion could well be produced by one and the same substance influencing membrane activity and thus the water metabolism and secretory activity of the cells. The only exceptions are neurosecretion from the lateral nsc, shown to be engaged in inducing circadian rhythms of activity and the effects of neurohormone C. The effect of cholesterol would be that of a vitamin supplying the steroid skeleton necessary for the production of ecdysone. This is necessarily lacking in decapitated or decerebrated insects unable to accept food. In such cases its supply by injection can reinduce the production of MH when at least a small amount of AH is present. But this, is not the case in late diapause when AH is absent.

Williams (1963, 1968), first thought that AH was hyaluronic acid or a related substance and later expressed the view that it might be a mucopolysaccharide. Neither of these hypotheses was confirmed by further investigations.

Contemporary research work by Gersch and his colleagues submitted definitive evidence in support of their original opinion. First of all they showed that the prothoracic gland stimulating agent is different from neuro- hormone D, previously described, and that it actually affects the prothoracic gland. Using electrophoretic separation on polyacrylamide gels, they further succeeded in demonstrating that the actual AH, i.e. the prothoracic gland-stimulating hormone consist of two components – one with high molecular weight, stimulating RNA synthesis, and the other, AH_{II}, with low molecular weight, which raises the membrane potential and hence membrane permeability. Both components are of a peptide nature. However, It was discovered that Neurohormone D, is a peptide with a molecular weight of about 2000, to be thermo- and acid labile and to be decomposed relatively quickly by trypsin.

The mode of action of AH

The movement of the neurosecretory granules through the axons of the nervi corporum cardiacarum from the nsc of the pars intercerebralis protocerebri into cc is well known, but their role in cc and their passage from there into the haemolymph is yet unknown.

Significant experimental evidence for the transport of AH from the cc to the pgl in *Rhodnius* by the haemocytes was produced by Wigglesworth (1956b). He showed that blocking the haemocytes by injecting Chinese ink, trypan blue or iron saccharate, the particles of which are phagocytosed, results in a significant delay of the next moult if the injection is carried out before the end of a specific critical period which corresponds approximately with that of AH. However, when at the same time he implanted active pgl or injected a sufficient amount of a solution of crystalline ecdysone, no such delay occurred. Wigglesworth concluded that under normal conditions the haemocytes transport AH from the cc to the pgl, or, it is possible, that they secrete a further substance necessary for the activation of the pgl under the influence of AH. As suggested by Wigglesworth, other explanation are so possible, for example, that AH is adsorbed on to the injected material and removed with the material from the haemo- lymph by phagocytosis; or the haemocytes, damaged by phagocytosis, discharge some AH inactivating substance into the haemolymph. An alternative explanation could be that, in normal insects, the role of the haemocytes is to phagocytose neurosecretory granules, thus freeing the AH whilst digesting the carrier substance. This would be in agreement with the observed occurrence of neurosecretory granules in the aorta dorsalis of various insects and with the observation of Hodgson and Geldiay (1959) who found that hyperactivity in Blaberus craniifer and, to a lesser extent, electrical shock treatments, resulted in bloods cells overflooding of all parts of the brain.

Hardly any information is one available regarding the mechanism of activation of pgl and other organs influenced by AH. But, it has been found by Williams (1952) that the same activating effect may by obtained by implanting another, active pgl. This is to be interpreted, together with Wigglesworth's (1957) conclusions, as meaning that continuous activity by AH is must for the pgl to provide an effective amount of moulting hormone. This is based on the discovery that the first change in the epidermis definitely attributed to pgl hormone commences about two days before the end of the critical period for AH, during which time the removal of the AH-source by decapitation stops the process of the moulting process.

THE MOULTING HORMONE (ECDYSONE) (MH)

(Growth and differentiation hormone, Scharrer, 1948; Williams, 1952; = Larvenhautungshormon = Verpuppungshormon = Imaginal-hautungshormon, Piepho, 1938; = Metamorphosehormon, Weber, 1954; = ecdysone, Butenandt and Karlson, 1954)

The initial indication of the existence of a humoral factor controlling moulting can be found in the experiments of Kopec (1917, 1922), and later Wigglesworth (1934). During time of Wigglesworth, Fraenkel (1934, 1935) showed that similar hormonal factor, necessary for puparium formation, was present in the brain region of *Calliphora erythrocephala larvae*. Prior to this, however, Hachlow (1931), on the basis of his experiments with butterfly pupae (*Vanessa io* and *Aporia crataegi*), suggested the existence of a 'thoracic centre' which was necessary for development. Bodenstein (1933a, b, 1934) in his transplantation experiments with the legs of caterpillars got the same result.

The honour of being the first person to clearly distinguish between the hormone of the brain cells and that of a new hormonal source, the prothoracic glands, belongs to Fukuda (1944) who showed the importance of pgl for moulting by transplantation experiments in silkworms (*Bombyx mori*). After this, papers on the moulting hormone appeared at an increasing rate. The role of pgl, or the analogous peritracheal glands in lower Diptera, or the pericardial glands or ventral head glands in Hemimetabola, in the moulting processes has been demonstrated in all the chief groups of insects.

Earlier the research recognized and were mainly concerned with the importance of MH for inducing the moulting process, but most of the later investigators stressted on its indispensability for growth and morphogenesis and used for it the less suitable terms 'growth and differentiation hormone', or 'metamorphosis hormone' (see above). A certain amount of confusion appears to have been caused by the discovery of the effects of the ring gland in flies, which is a composite structure that contains the sources of all three metamorphosis hormones also, the role of MH in diapause and imaginal differentiation in *Cecropia* pupae as resolved by Williams seemed at first to supportthe concept of the growth and differentiation concept.

The most significant stage in MH invitations, after its separation from AH by Fukuda, was its isolation in crystalline form by Butenandt and Karlson (1954) from extracts of silkworm pupae. A significant requirement for this was the discovery by Becker and Plagge (1939) of a suitable test organism and a quantitative measure for judging concentration, the so-called *Calliphora*-unit. This method was improved by Butenandt and Karlson in their isolation experiments. While the specificity claimed by Williams (1951a, b) for the so-called spermatocyte-test appears to be questionable in the light of the findings of Laufer (1960). Becker and Plagge (1939) were also the first to show the broad inter-Order non-specificity of MH, which was confirmed by Wigglesworth for such widely separated orders as Hemiptera and Diptera. Another significant step was the successful extirpation of the ventral glands in migratory locusts by P. Joly, L. Joly and Halbwachs (1956).

The theory of the pgl hormone as a MH, which corroborates with the original findings of Kopec and Wigglesworth (1934) as well as with the gradient-factor theory of the author, has been approved by two independent pieces of work: Halbwachs and Joly (1957) showed, using *Locusta migratoria*, that transplantation of the pgl accelerates the moulting process without utilizing any positive effect on growth differentiation; Luscher and Karlso (1958) observed moulting but no and growth or imaginal differentiation following the injection of a large quantity of ecdysone into the nymph of *Kalotermes flavicollis*.

These discoveries have been confirmed by many other researchers For instance, Zdarek and Slama (1972) proved that the injection of a large dose of ecdysterone at the outset of the last larval instar in *Calliphora* and *Sarcophaga*, inhibits morphogenesis and results in a supernumerary larva, whereas small amounts at later stages simply depress growth and result in small puparia. Administration at interval times produce forms between the larva and pupa.

The impact of ecdysone on colour change in *Cerura vinula* caterpillars (Buckmann, 110) also conforms with this theory; the greater the dose of hormone injected the quicker becomes the process of moulting whilst the effect on colour change decreases with an increase in the amount of hormone.

The function of MH (together with that of JH) at the cellular level was examined in detail by Wigglesworth (1963c) in the epidermal cells of *Rhodnius prolixus*. He came to the conclusion that MH is not a necessary ingredient for growth in insects in general, but it is necessary for the activation of the epidermal cells to produce their secretion (the moulting fluid and the chitinous cuticle) and to grow and divide. But precisely the same response is obtained by 'wound hormones' from injured tissue (cf. Wigglesworth, 1937). And the cells of the fat body and haemocytes do not need MH at all – their activation and mitotic activity is brought about by nutrition alone. The first discernible effects of MH on the nucleolus of the epidermal cells is discussed and contrasted with the effects on the puffing patterns in the salivary glands in *Chironomus*.

Signitcant evidence supporting the view that MH only indirectly stimulates growth and morphogenesis is provided by the distribution of DNA synthesis during insect development. These authors conclude that ecdysone should be viewed primarily as a moulting hormone that initiates biosynthetic activities which caused moulting.

In some tissues, such as all chitinogenous epithelia, the Malpighian tubules and the nervous system, DNA synthesis occurred soon after MH secretion started. In others, e.g. pgl, midgut and haemocytes, it continued throughout the whole of the larval moulting cycle, but acquired a maximum at the peak of ecdysone production. In further tissues, e.g. the imaginal wing discs and muscles, no correlation between DNA and ecdysone production was discoverd.

Schaller and Andries (1970a, b,) got the same reults while studying regeneration and metamorphosis of the midgut cells in *Aeschna cyanea* and by Mouze and Schaller (1971) and Mouze (1971) in a study of development of the eyes in the same species, based on observations of mitotic activity. Agui *et al*. (1972) showed that an explanted pgl of *Mamestra brassicae* clearly stimulated the induction of moulting in the integument of the diapausing Stemborer (*Chilo suppressalis) in vitro*. On the other hand, MH failed to stimulate spermatogenesis in naked cysts of *Cecropia* pupae, but was active in a culture of intact testes.

Morohoshi and Iijima (1969) and Morohoshi *et al*. (1972) also reached the same conclusions with larvae and pupae of *Bombyx mori*. The only contradictory evidence is the report by Sondhi (1968), who claims to have found that in *Drosophila melanogaster* the injection of active ring glands into inbred larvae of the same sex and age did not affect either the time of puparium formation or of adult emergence, but raised the recipients' wet and dry weight by up to 10 per cent. The rate of the increase and the number of individuals are too few to allow definitive conclusions to beobtained from the results.

MH has a primary, direct effect on tissues of ectodermal origin only, i.e., the epidermis, stomodeum and proctodeum, the epithelia of the tracheal system, the ectodermal parts of the imaginal discs and the nervous system, etc. Its presence in the minimum active concentration, in the absence of JH, determines growth of the imaginal parts and degeneration of the larval parts according to the species specific morphogenetic pattern. This, together with induction of the moulting process, may *per se* be enough indirect stimulus to initiate metabolic and growth activity in other, MH-independent, tissues also

The apolytic glands (agl)

Many glands in the insect body have been considered to produce MH. They have some common characteristics. They are paired, laterally localized glands, mostly ribbon-like in form, and sometimes more or less branched. Their main component is large parenchymal cells with 'intricately interwoven cytoplasmic processes extending towards the glandular surface' (Scharrer, B., 1964). where their ultrastructure has been examined, different quantities of smooth tubuli of the endoplasmic reticulum has been found. These are supposed to be connected with steroid production. The glands also seem to have the same embryonic and phylogenetic origin.

There are six classifications of these glands (see below) have been described in different parts of the body in various groups of insects.. Perhaps they can be extended to include the larval oenocytes. It was felt that a common term, applicable to all these glands, was needed and some authors suggested the collective designation 'ecdysial glands'. Wigglesworth (1962) however, pointed out that this term has already been applied to a certain type of epidermal gland and might thus cause confusion. Another contrary reason is that ecdysis is precisely that part of the moulting process which is uncontrolled by the hormone from these glands. I would therefore suggest calling them 'apolytic glands' and have employed this common term in the present book. It is derived from 'apolysic', the first stage in the moulting process under the direct control of the glands in question.

Many researchers has expressed their doubts (Locke, 1969; Romer, 1971, as to whether the agl actually produce MH. But these doubts seem to be vitiated by the findings of Agui *et al*. (1972) and Kambysellis and Williams (1972), who have demonstrated the activity of the glands *in vitro.* This discovery is substantiated by numerous authors in vast quantities of early and recent experimental data, whereas the contradictory observation can be explained by assuming that a primary MH source does exist.

1. The prothoracic glands (pgl)

Morphology

The pgl of the *Cossus cossus* caterpillar were almost accurately described by Lyonett as early as 1762. A more complete description

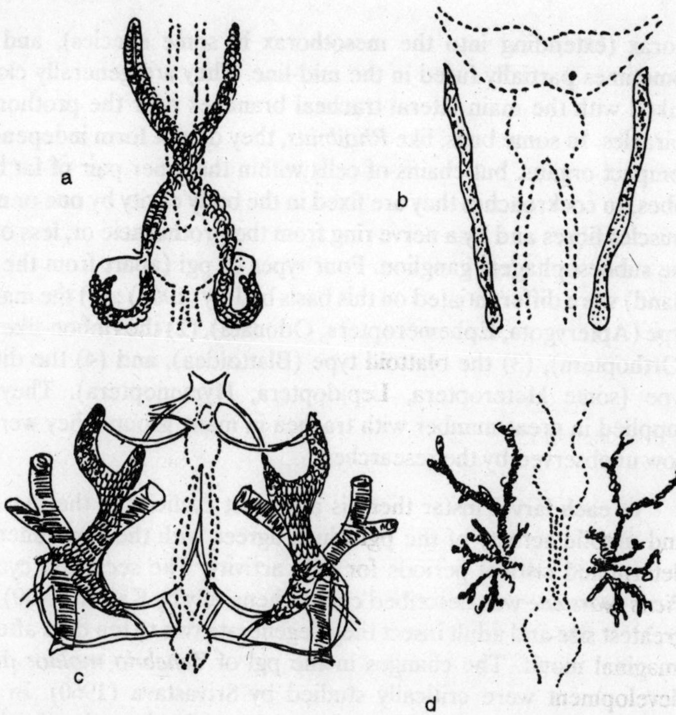

Fig. 7.7. Prothoracic glands in four different insect orders : *(a)* Blattoptera, *(b)* Hemiptera, *(c)* Lepidoptera, *(d)* Hymenotpera, diagrammatic, after various authors modified.

along with data on embryonic development is given by Toyama (1902) who called them 'hypostigmatic glands'. Ke (1930) used the term pgl for the first time. Ever since because of the interest exicted by the discovery of their function, they have been described in detail for all the important groups in insects.

Many comparative morphological studies of the pgl and analogous organs in various groups of insects have also appeared. For example, they were described by Lee (1948) in many lepidopterous larvae, by Pflugfelder (1947d) in various orders of Hemimetabola, by M. Thomsen (1951) in Diptera, by Wells (1954) in Hemiptera, by Srivastava (1959) in Coleoptera, etc.

Although they change a great deal in shape, the pgl have certain features in common in all groups where they occur : they are paired glandular structures located in the ventrolateral areas of the prothorax (extending into the mesothorax in some species), and are sometimes partially fused in the mid-line. They are generally closely linked with the main lateral tracheal branches near the prothoracic spiracles. In some bugs, like *Rhodnius*, they do not form independent compact organs, but chains of cells within the inner pair of fat body lobes. In cockroaches they are fixed in the body cavity by one or more muscles fibres and by a nerve ring from the prothoracic or, less often, the suboesophageal ganglion. Four types of pgl (apart from the ring gland) were differentiated on this basis by Joly (1968) : (1) the massive type (Apterygota, Ephemeroptera, Odonata), (2) the ribbon-like type (Orthoptera), (3) the blattoid type (Blattoidea), and (4) the diffuse type (some Heteroptera, Lepidoptera, Hymenoptera). They are supplied in great number with trachea in many groups they were till now unobserved by the researchers

In each larval instar there is a distinct cyclicity in the secretory and mitotic activity of the pgl which agrees with the experimentally determined critical periods for MH activity. The secretion cycle in *Pieris brassicae* was described comprehensively by Kaiser (1949). The greatest size and adult insect they degenerate two to ten days after the imaginal moult. The changes in the pgl of *Tenebrio molitor* during development were critically studied by Srivastava (1960). In each larval instar they reach their maximum size at the time when feeding is interrupted, i.e., about two days before ecdysis (cf. the critical period for MH); they produce most of their secretion at this time and afterwards become reduced. A new cycle begins with the feeding of the next instar. The critical period occurs less than 24 hours after the interruption of feeding in the last larval instar. The glands are completely reduced at the commencement of eye pigmentation. However, they persist throughout the life of insect in Apterygota.

Herman and Gilbert (1966), accomplished a detailed anatomical and histological study of the pgl in *Hyalophora cecropia*, and found numerous, very diffusely arranged chains of cells, reaching to the posterior portion of the thorax, which arose from the four branches (anterior, dorsal, ventral and posterior) of the relatively compact tissue of the gland round the first spiracle. Each gland contained some 200 to 290 cells, which were larger and usually more in number in females and degenerated immediately after adult emergence.

Histology

The pgl are usually formed of two strips of glandular tissue. The essential components of these are glandular cells equal to those of the corpora allata. Their cytoplasm is basophilic staining deeply with methylene blue, neutral red and other stains. Numerous deeply staining granules are found in the cytoplasm. Abundant black granules, probably lipoid in character, are found after fixation with osmium tetroxide. The glandular cells are joined by intercellular bridges which stain blue with azan and are connected by their anastomoses with the membrane enveloping the axial muscle fibres. A rich supply of thin tracheae was observed by Wigglesworth (1952a) in *Rhodnius prolixus* in contrast to the feeble tracheation of the surrounding fat body. In contrast to the pgl in cockroaches and butterflies, no nerve fibres were observed in the pgl of *Rhodnius*, either in dissected glands or histologically.

In *Hyalophora* each pgl cell consists of peripheral with striated border through which the secretory substance is released. Cycles of activity are evident, characterized by nuclear changes followed by cytoplasmic vacuolization and correlated with the moulting cycle. A low level of secretory activity is discovered in young pupae and a higher one in older pupae. Discernible variability of secretory activity could be seen among the cells of the same gland. A detailed histological study of the pgl of *Galleria mellonella* was carried out by Mala *et al.* (1972) with the aim of investigating JH effects at both the histological and the ultrastructural level. The more compact gland is composed of about 55 polyploid cells with giant nuclei and different layers of cytoplasm. The secretory cycles connected with the moulting process were examined comprehensively in the penultimate VIth and last VIIth larval instar and the pupal instar and differences between them were ascertained

The amount and distribution of DNA and RNA in the pgl of *Samia cynthia* and several related species was considered by Oberlander *et al.* (1965) in the penultimate and last larval instars and in pupae. DNA synthesis was found to be quite high in the IVth larval instar and lower in the last larval instar and pre-pupa. No DNA synthesis was observed in the pupa during diapause or during adult development. In these two stages, even JH injection failed to induce synthesis. How- ever, RNA synthesis was several times higher during diapause than during adult develop- ment, and was easily stimulated by JH injection. Gersch and Sturzebecher (1970) observed a similar increase in RNA synthesis in the pgl of *Periplaneta americana* nymphs following induction by AH. Eight hours after injecting the hormone, a noteworthy increase in the mixture of stearic, oleic, linoleic, linolenic, palmitic and myristic acids, but not MH (ecdysone) was detected in the glands.

The phases of pgl secretory activity were examined histologically and ultrastructurally by Hintze (1968) in *Cerura vinula*, in relation to the corresponding cycles of the nsc of the brain, ca and the dorsal thoracic gland. The characteristics of the pgl in the active state at both the light and electron microscope levels are described and the penetration of haemocytes across the tunica propria into the pericellular space is reported. The innervation of the pgl and the possibility that brain neurosecretion has been transported to the glands along these axons is discussed.

Ultrastructure

The inital electron microscopy study of insect pgl was accomplished out by B. Scharrer (1966) with cockroach *(Leucophaea maderae* and *Blaberus craniifer*) nymphs at various stages of the moulting cycle. She considered the following to be their most distinctive cytological features : (1) long and intricately interwoven parenchymal cell processes, the cells sometimes being separated from each other by extracellular channels with an average width of 0.5 μ; (2) a specific surface membrane (external lamina) of extracellular origin; (3) numerous micropycnotic caveolae and vesicles (thought to facilitate the transport of material from the glandular cells to the haemolymph or vice versa) covering a large area of the cell surface; (4) Clear but never visible Golgi units; (5) a small amount of smooth endo- plasmic reticulum characteristic of steroid-producing cells, sometimes with a striking array of micro-tubules; (6) a varying incidence (according to the phase of the moulting process) of other structures, such as nucleoli, mitochondria and lysosomes. Degeneration of the glands started within a few days after adult ecdysis. Early signs of their destruction were observed in the nuclei, in the form of patches of contrasting electron density characteristic for pycnosis. Later, large heterogeneous inclusion bodies, thought to be autophagic vacuoles, appeared in the cytoplasm. In the advanced stages of degeneration, the plasma membrane and nuclear membrane

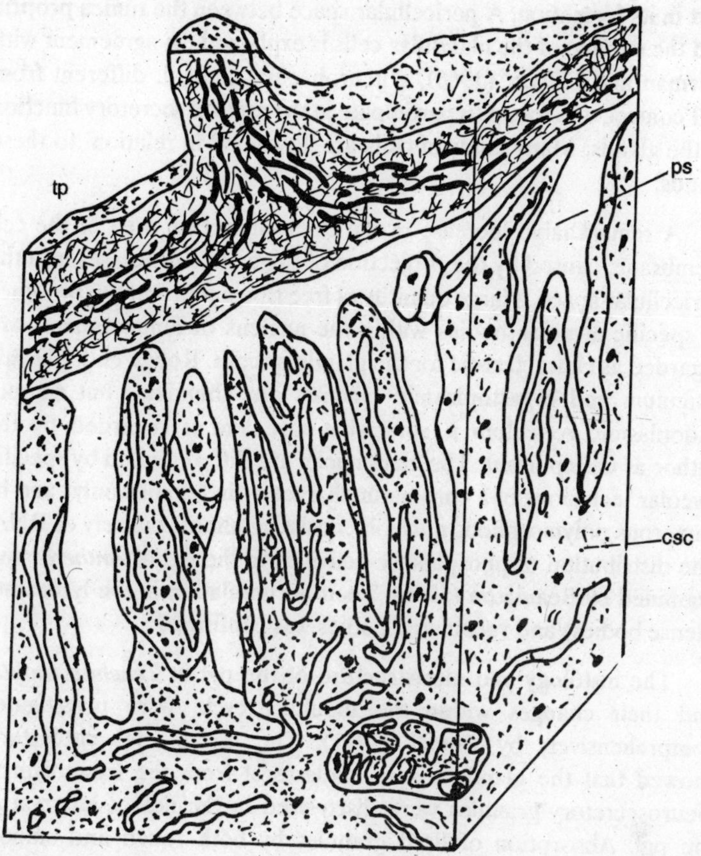

Fig. 7.8. Block diagram of the surface layer of a prothoracic gland cell of *Antherea pernyi* as seen under the electron microscope. *tp*—tunica propria, *ps*—pericellular space, —cytoplasm of secretory cell.

were slowly autolysed and the protoplasm was invaded by phagocytic haemocytes.

An exhaustive electron microscopical study of pgl in Lepidoptera (*Bombyx mori, Antherea pernyi* and certain other Saturniidae) was carried out by Beaulaton. The structure and origin of the tunica propria, the pgl surface membrane, wa sinvesti- gated. It is perhaps matches with the basement mem- brane of the epidermis , since it has a similar microfibrillar structure and several types of haemocytes take part in its formation. A pericellular space between the tunica propria and the surface of the glandular cells is explained. In agreement with Herman and Gilbert (1966), a neurohaemal control, different from AH control, was found to participate in the cyclical secretory function of the glands. Many kinds of cells are described in relation to these glands.

A remarkaable increase in the external surface area of the cell membrane, caused by the projection of finger-like processes into the pericellular space, a large quantity of free ribosomes and the presence of specific alveolar bodies within the nucleus of female glands are regarded as characteristic for the glandular cells. Rough endoplasmic reticulum and ergastoplasmic saccules are abundant, but smooth endoplasmic reticulum is somewhat rare and is regarded by the author as unimportant. The giant nuclei are differentiated by specific alveolar inclusions of a chromatin character (in females only) and by numerous polymorphous nucleoli consisting almost entirely of RNA. The distribution of phosphatase activities in the pgl of *Antherea* was examined by Beaulaton (1966). Their accumulation in the lysosomes (dense bodies) and vacuolated bodies was confirmed.

The histology and ultrastructure of the pgl in *Tenebrio molitor* and their changes within the moulting cycle were investigated comprehensively by Romer (1971). The use of ^{3}H–cholesterol showed that the glands absorb cholesterol from the haemolymph. Neurosecretory granules were discovered in the nerves innervating the pgl. Absorption of lipid granules by both rough and smooth tubules of the endoplasmic reti- culum was observed. The extensive Golgi system was expected to be concerned with the production of ecdysone from cholesterol. Extracts of isolated pgl were shown to invite pupation in *Calliphora*.

Mala *et al.* (1974), Novak *et al.* (1974) and Blazsek *et al.* (1975) made a detailed study of the structure and ultra- structure of the pgl in the last two larval instars and pupal instars of *Galleria mellonella*, for the purpose of determining JH effects. According to them all structural differences between the penultimate and the last larval instar as being due to JH deficiency in the latter. In consequence, the cytoplasm is vacuolated and less compact, and at first an increase in the amount of secretion is effected in the gland.

The impact of JH and MH on nuclear RNA synthesis in the pgl and corpora allata of saturniid pupae *(Philosamia cynthia* and *Hyalophora cecropia*) were examined in an autoradiographic study by Siew and Gilbert (1971). The pgl were stimulated by MH within three hours, ca three hours later. JH also stimulated both glands, but in this case the ca were activated sooner than the pgl. The adminis- tration of MH to pupae in which the adult moulting process had just started, resulted in drastic reduction of nuclear RNA synthesis. This is in agreement with the findings of Socha on hormonal interactions in various phases of the isntar.

Embryogenesis

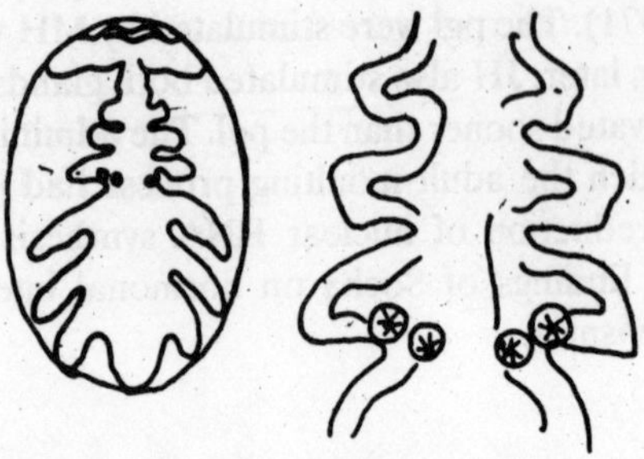

It has already been demonstrated by Toyama (1902) that pgl appear at a very early stage in embryonic development as an epidermal invagination of the lateral portion of maxillary segment and from here later extend into the prothorax and, where this is reduced, even into the mesothorax such result was also obtained by Wells (1954) for the bug Dysdercus cingulatus. Here, two pair of invaginations arise in the second maxillary segment; the anterior, larger, pair develops into labial glands whilst the posterior pair gives rise to the prothoractic glands (Fig. 14). If, however, their suppposed origin form the nephridian tubules and their innervation (in most insects by nerves from the prothoracic ganglion) is taken into account, it appers more probable that the invagination mentioned does not start in the second maxillary segment but in the most anterior portion of the prothoracic segment.

The figure given by Wells (p. 91, Fig. 16, cf. Fig. 14 of the present work) does not exclude the possibility: 'It is quite possible that in certain insects the pgl may be shown to be the derivatives of nephridia of the girst thoracic segment' (Scharrer, B,1948).

Phylogenesis

Pgl, or thier equivalents are today known in practically all groups of metabolous insects (Pterygota). The conclusion, however, that their occurrence is not restricted to the Pterygota is well founded. Pfluggelder (1958) suggested the possibility of a homology between the ventral glands of Hemimetabola and the so-called head nephridia of Apterygotawhich are aslo developed from the second maxillary

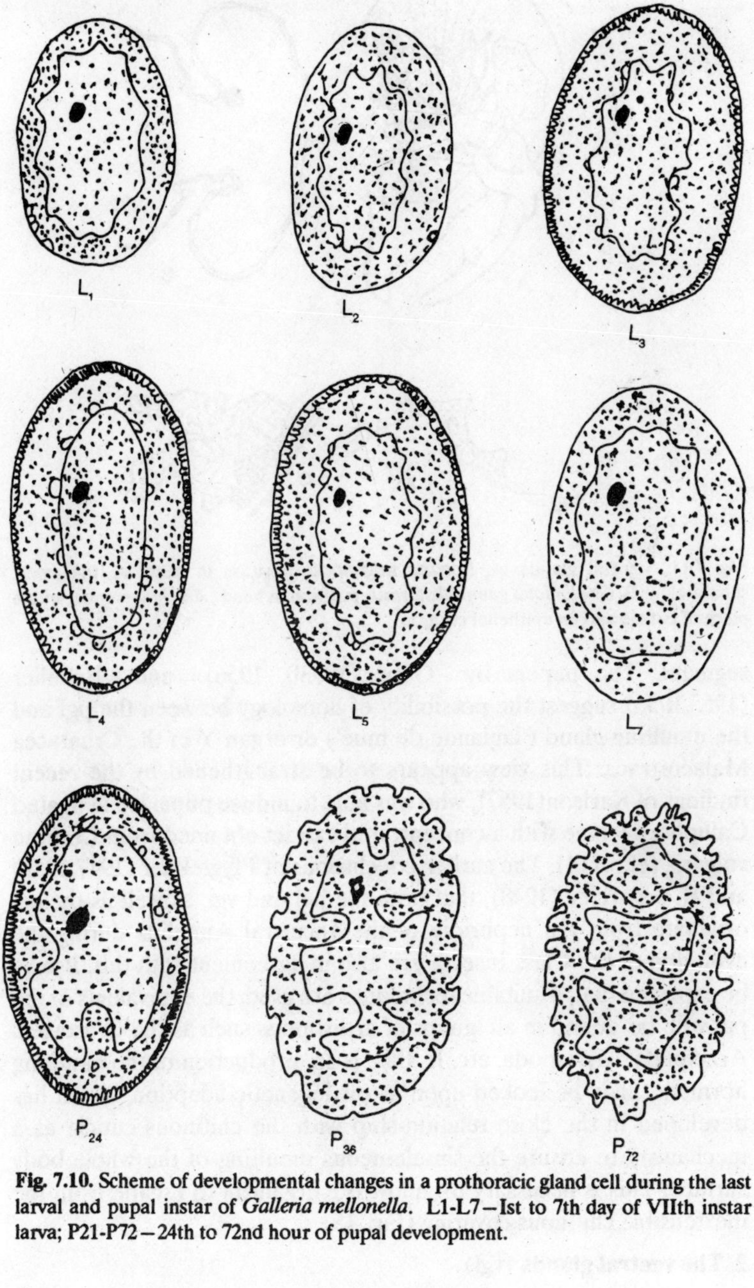

Fig. 7.10. Scheme of developmental changes in a prothoracic gland cell during the last larval and pupal instar of *Galleria mellonella*. L1-L7 – Ist to 7th day of VIIth instar larva; P21-P72 – 24th to 72nd hour of pupal development.

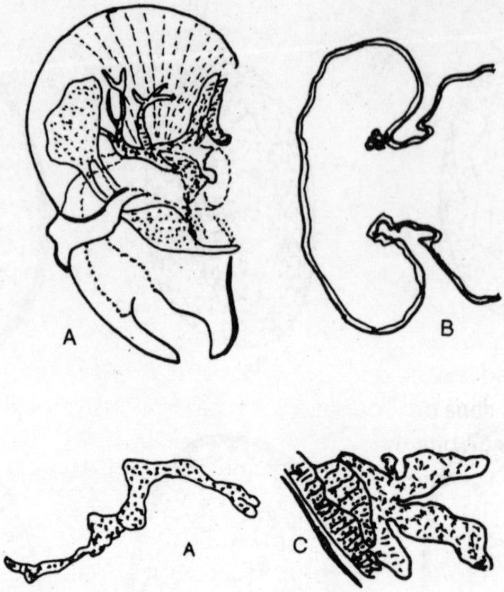

Fig. 7.11. Ventral glands. A, *Locusta migratoria*, position in head; A, the same, separately); B, *Grylloblatta campodeiformis*, position in head ; *Aeschna cyanea*, with a piece of cuticle and an epithelial cushion

segment. The papers by Gabe (1953b, 1956) and Echalier (1955,1956) suggest the possibility of homology between the pgl and the moulting gland ('laglande de mue') or organ Y in the Cruatacea Malacostraca. This view appears to be strangthened by the recent findings of Karlson(1957), who was able to induce puparion in ligated Calliphora larvae with a concentrated extract obtained from Cranton vulgaris (cf.p.114). The earlier conclusions of Pfygfekder (1947,1952) and B. Scgarrrer (1948), that both the pgl and vgi, as well as the ca, originate from the nephridia of the ancestral Annelids during the evolution of the Class Insecta are also in agreement with this. It may be expected that, if suitable techniques are used, the equivalents of the pgl will be found in all goups of arthropods such as the Crustacea. Arachnida. Myriapoda, etc. If this is so, production of the moulting hormone may be looked upon as phylogenetic adaption, which has developed in the close relation-ship with the chitinous cuticle as a mechanism to ensure the simultaneous moulting of the whole body surface, This is necessary to allow the organism to escape from the inextensible chitinous covering (Fig. 15).

2. The ventral glands (vgl)

Morphology

Vgl, often called ventral head glands or tentorial glands, were first found by Pflugfelder (1947a), in the hind ventra region of the head in Phasmidae. Boisson (1947) described corresponding structures in *Bacillus rossii* as corpora incerta and discovered a cuyclicity in their secretory activity, which was connected with moulting, Pflugfeelder (1947d) assumed that they were absent in Hemiptera and in all Holometanola.

An important feature of the vgl is that they degenetate at the end of metamorphosis or in the first few days after the imaginal moult inthe same way as the prothoracic glands (= tentorial glands) of the worker and soldier castes of termites., which are less developed than in the sexuales but are active during the whole life period (Springhetti and Bernardini, 1955; Kaiser, 1956). This appears to be connected with the neotenic character of these castes insome lower termite groups, evidenced by the underdeveloped state of their ovaries, in which they are maintained by inhibitory substances (cf.p. 415 and Fig. 16) produced by the reproductives (cf.Pflugfelder, 1958).

The influence of the vgl on the moulting process was shown by Strich-Halbwwachs (1954, 1958) and by Halbwachs, Joly and Joly (1957) in migratory locusts. Tyhe implantation of an active gland at the geginning of the IVth or Vth instar causes an acceleration of the moulting process with a feeble prothetelic effect (cf. p. 99). The so-called corpora adnexa in *Dinjapyx marcusi* which are epithelial in structure and situated caudally to the suboeso phageal ganglion are viewed by Marcus (1951) as a homologue of the vgl in Apterygota. Some authors suggested that the vgl can be understood as homologous with the pgl of Holometabola. But, it seems more probable that they are serial homologues of the pgl .

A regular cyclical secretory activity of vgl in *Carausius* was described by Pflugfelder (1958). The drops of secretion appear, closely connected with the nuclei, during and for a few days after ecdysis. This period of excessive secretion is followed by a phase of exhaustion. Between the 3rd and 7th day of the intermoult period both the nuclei and cytoplasm grow rapidly. On the 8t and 9th days, there are many cell dimensions resulting in abundant cell aggregates between the 10th and 11th days. Ecdysis takes place on the 12th day r relatted with the strong secretory activity. Similar observations have been made in Odonata, termites and other Hemimetabola.

Histology and embryology

The histological structure of the vgl is quite similar to that of the pgl. Their embryogenesis was examined by Pflugfelder (1958) in *Carausius morosus*. They develop from epidermal proli- ferations in the ventrocaudal part of the head, but are soon separated from the epidermis and become bladder-like. Jones (1953, 1956) found vgl in *Locusta pardalina* and *Locusta migratoria* as invaginations in the head region of embryos at the end of the katatrepsis stage. He considered them to be homologous with the pgl of other insects. He examined their function in relation to the beginning of the secretory activity of the neurosecretory brain cells and the first embryonig moulting process.

Degeneration of the vgl of the earwig *Anisolabis maritima* during metamorphosis was studied histologically and analysed experi- mentally by Ozeki (1968, 1970), who transplanted them to various other stages of the same species. He concluded that both the small amount of JH in the last larval instar and the latter's internal state were determining factors in the regression of the vgl. The implantation of active vgl from larvae to adults led to a supernumerary moult in the adult stage

Ultrastructure

Joly *et al.* (1969) studied the vgl of *Locusta migratoria* using the electron microscope. The major structures found in the cytoplasm were the Golgi apparatus, ergastoplasm and structures of lysosomal or autophagic character; no smooth endoplasmic reticulum was observed. The authors concluded that a substance which was probably protein nature, and not MH, was produced. In this they agree with Locke (1969,) the vgl produces a protein substance, which merely controls ecdysone production elsewhere in the body. Joly *et al.* (1969) described special structures consisting of degenerating rough endoplasmic reticulum. In another research paper from the same laboratory, published soon afterwards the authors formed a different opinion, as in the oenocytes they found an abun- dance of smooth endoplasmic reticulum, which may be related with the occurrence of a steroid component in the cuticular lipids. Unlike Locke (1969), they did not consider cytological structure alone as correct evidence for steroid production.

In an ultrastructural study of the secretory cycle in *Locusta migratoria* vgl, Cassier and Fain-Maurel (1968) based their conclusions on the belief that the glands participate in ecdysone production. They demonstrated that, during the vgl activity, cycles the tubules of the smooth endoplasmic reticulum were extruded from finger-like cytoplasmic evaginations into the intercellular spaces of the gland. They stressed that this extrusion coincides with maximum MH activity and concluded that it was the most effective way of discharging the hormone from the glandular cell. They suggested the use of isotopically-labelled cholesterol in order to solve this question.

Differences in the ultrastructure of vgl in the active state and during diapause were examined by Guellin (1971) in the grasshopper

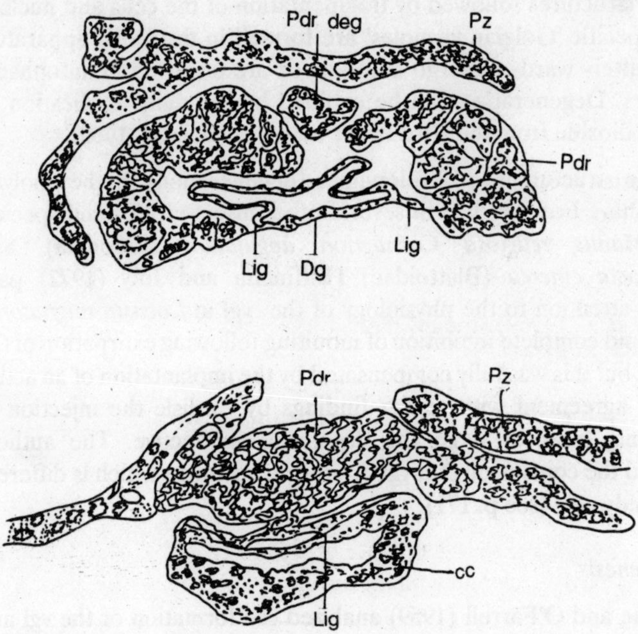

Fig. 7.12. Pericardial glands of *Phyllium* sp. in transverse sections. *Pdr*—pericardial glands; *Pdr. deg*—degenerated part of a pericardial gland; *Pz*— pericardial cells; *Dg*—dorsal vessel; *Cc*—corpora cardiaca; *Lig*—Ligaments.

Pyrgomorpha conica. The ultrastructure of diapausing glands is indicative of remarkable depression of metabolism : the pericellular system is reduced, lysosomes are apparently absent and free ribosomes and ergastoplasm are rare. From the development of the ergastoplasm at the end of diapause, the author concludes that active vgl produce a protein type of substance, as did Joly *et al*. (1969) in *Locusta*.

Fain-Maurel and Cassier (1968, 1969) examined the involution of the vgl in the last larval instar of the gregarious phase of *Locusta migratoria migratorioides* (in the adult, the glands persist in the solitary phase) and compared it with the secretory cycle in earlier instars. They claim 'an early and primordial role of the Golgi apparatus; in the elaboration of the 'vacuolar bodies' of 'cytosomes' and participation of rough endoplasmic reticulum in the process of degeneration, through the formation of 'cytosegresomes' and 'particular microtubular areas'. The process begins with the autolysis of the named structures followed by fragmentation of the cells and nuclear lysis. Specific 'Golgian vacuoles' are formed in the Golgi apparatus, which after- wards undergo autolysis or are enclosed in autophagic vacuoles. Degeneration can be avoided by the daily application of carbon dioxide from the IIIrd instar onwards (Baudry, 1969).

The structure and development (degeneration) of the apolytic glands have been studied in several other insect groups and species, e.g., *Mantis religiosa Coenagrion angulatum* (Odonata), and *Nauphoeta cinerea* (Blattoidae) Hoffmann and Joly (1972) paid special attention to the physiology of the vgl in *Locusta migrtatoria*, and found complete inhibition of moulting following extirpation of the glands, but this was fully compensated by the implantation of an active pgl. In agreement with earlier findings by Carlisle the injection of ecdysone, however large the dose, was ineffective. The authors reached the conclusion that vgl produce a hormone which is different from ecdysone (see p. 171).

Phylogenesis

Rae and O'Farrell (1959) analyzed the formation of the vgl and the retrocerebral complex in *Grylloblatta campodeiformis*, a represent- ative of the primitive Orthopteroid Order Notoptera (= Grylloblat- tidae). They explained the structures obviously endocrine in function in the ventrocaudal region of the head and in the cervical area which resembled very closely the so-called 'head lobes' of *Blatella germanica* and the 'cervical glands' of *Periplaneta americana*. These researchers accepted that the coxal muscles they describe in *Grylloblatta* are the remnants of the degenerated pgl of cockroaches, and homologous with the axial muscle cord in Blattidae depicted as Scharrer's organ by Chadwick (1955).

The presence of both pgl and homologous of vgl in the cervical structures of cockroaches seems to be well accounted for by the morphological and experimental evidence of Rae (1955) and Chadwick (1955, 1956). Thus complete etirpation of the pgl of *Periplaneta americana* does not prevent moulting. Cyclical changes in the volume of the 'ventral lobes' in *Blatella* are related to the moulting process and regeneration

3. The pericardial glands

The initial indication of the existence of ductless glands associated with the dorsal vessel was made by Verson (1911a, b). The pericardial glands in various species of Phasmidae *(Phyllium, Carausius* etc. Were fully explained by Pflugfelder (1938a-d, 1949b). Besides the vgl they occur in the form of paired glands, the histological structure and function of which, unlike their origin, is equal to that of the pgl and vgl. They also decline in the adult.

4. The peritracheal glands

Matching structures were described by Possompes (1949 a-c, 1953a, b) and by other researchers in the Chironomidae, Simuliidae, Tabanidae, and other groups of Diptera under the name of peritracheal glands. They have not been discovered in Tipulidae. Both reseatchers agree their homology with the pericardial glands of Phasmidae. The relation of these to the moulting process was demonstrated experimentally by Pflugfelder (1949b). No moulting was observed in specimens in which the glands were extirpated.

Histology, embryology and phylogenesis

In their histological structure the pgl of both Phasmidae and Diptera correspond closely with the pgl and vgl of other groups of insects. The pericardial glands were initially accepted to be of

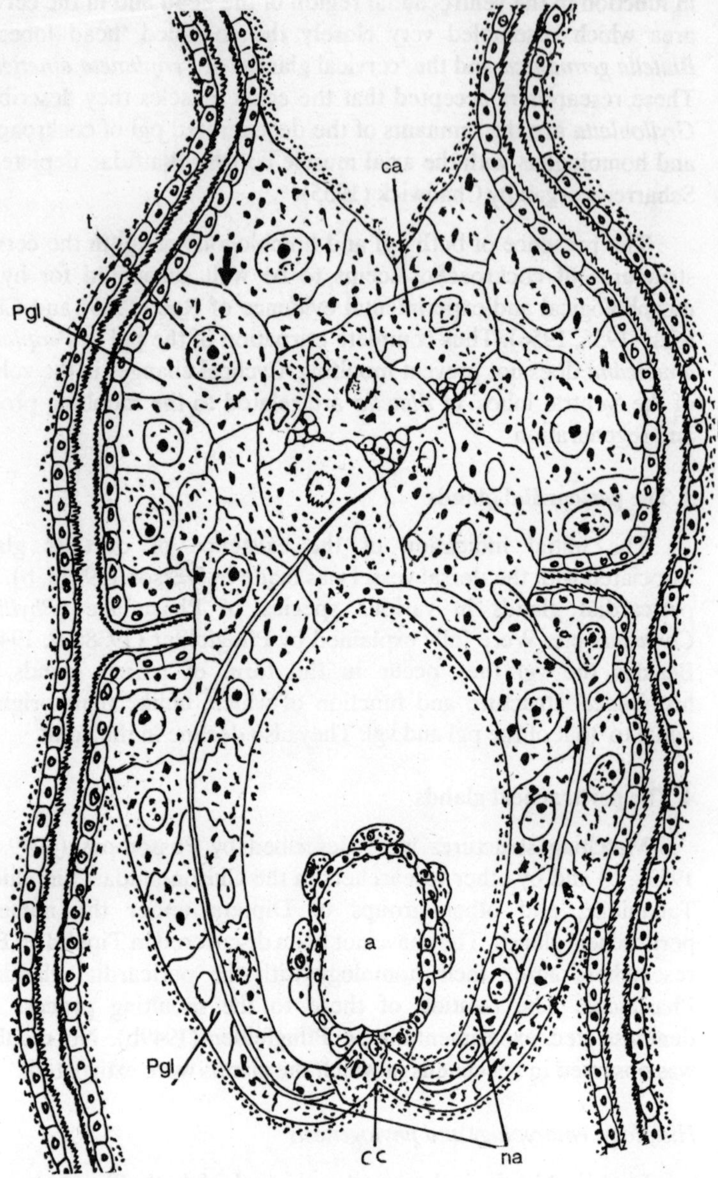

Fig. 7.13. Cross-section of ring gland of *Drosophila melanogaster* IIIrd instar larva. *ca*—corpus allatum, *t*—trachea, *pgl*— peritracheal gland cells, *a*—aorta, *na*—nervus allatus axons, *cc*—corpus cardiacum.

mesodermal origin by Pflugfelder (1938). In a later work however, Pflugfelder (1958) mentions their ectodermal-like structure and emphasized the difficulty in distinguishing between mesenchymal and ectodermal cells in the early embryonic period of *Dixippus morosus*. This would remove the only possible objection to their identification with the peritracheal cells, which are of ectodermal origin according to Possompes. The problem of their homology has, however, been rather complex because of by their confusion with pericardial cells of obvious mesodermal origin, which is not at all simmilar to the glands.

5. The ring gland (rgl)

In higher diptera special endocrine organ occurs, which combines the source of all three metamorphosis hormones, the first signs of which can be seen in the Tipulidae. It was initially explained by Weismann (1864) in *Calliphora vomitoria* and is therefore referred to by some authors as 'Weismann's ring'. The first to suggest its endocrine character was Hadorn (1937a-c) although the experiments of Fraenkel (1934, 1935). In these papers the rgl was expected to exert a positive influence on metamorphosis. But Burtt (1937, 1938) postulated that the rgl produced an opposite, i.e., inhibitory, effect on metamorphosis, and he identified the rgl with the ca. Hanstrom, on the other hand, attempted to equate the rgl with the cc on the basis that nerve cells are found therein, which do not exist in other endocrine glands.

The first suggestion made by B. Scharrer and Hadorn (1938) who first suggested the ring gland as a composite structure, consisting of both ca and cc. The structure of the rgl during metamorphosis and the development of the adult cc and ca in *Calliphora* were thoroughly studied by E. Thomsen (1941, 1947). She recognized the lateral portions of the rgl, formed of large glandular cells, with the pericardial glands of Phasmidae by showing that they disintegrate during metamorphosis. She also found that the hypocerebral ganglion is associated with this composite structure. Her findings were confirmed and elaborated by Vogt (1941c, d) for *Drosophila*, M. Thomsen (1951) for many species of Diptera, and by Possompes (1953) in a monograph chiefly concerned with the structure and function of the rgl in *Calliphora erythrocephala* based on the individual implantation of parts of the ring.

Four major components of the rgl may now be differentiated : (1) a corpus allatum of the small cell type situated in the dorso-median part of the ring, often covered by large cells of the lateral portions; (2) pericardial (peritracheal) glands formed by the large glandular cells of the lateral portions of the ring; (3) a corpus cardiacum in the ventromedian part of the ring, with only one a couple of nervi corporum cardiacarum formed by the coalescence of the nervi corporum cardiacarum interni and externi before they enter the ring gland fusing with the nerve recurrens in some species; (4) the hypocerebral ganglion consists of transparent ganglionic cells and closely joined with the corpus cardiacum.

Ultrastructure

An exhaustive study of the ultrastructure of the rgl of a normal and a 1/2/gl mutant strain of *Drosophila melanogaster* was carried out by King *et al*. (1966) and Aggarwall and King (1969). The quantity of smooth surface endoplasmic reticulum in the peritracheal gland cells increased about tenfold in the pre-pupal period in normal larvae, but did not exceed 1 per cent in mutant larvae. Metamorphosis of homozygous larvae of the lethal mutant was impossible unless an active gland from a normal larva was implanted. The researchers reached the conclusion that the relevant gene (the + allele of the 1/2/gl mutant) was indirectly concerned with the synthesis of smooth endoplasmic reticulum, where cholesterol from the food is transformed to ecdysone. In the intracellular spaces between the gland cells they encountered numerous cytoplasmic vesicles, which were pinched off from the cell projections.

Willig *et al.* (1971) examined ecdysone and ecdysterone biosynthesis in 7-day *Calliphora erythrocephala* larvae by injecting the larvae with ^{14}C-Cholesterol. The same kind of experiment was performed with explanted brain-rgl complexes *in vitro*. In the latter case, no active hormones were found, but occasionally a very small amount of active ecdysone glycosides or esters were produced, which exhibited lot of hormonal activity after enzymic hydrolysis with α-glucosidase of esterase. Therefore, the researcheres regard the inactive hormone glyco- sides as MH precursors produced by the gland.

The ultrastructure of the rgl of *Drosophila melanogaster* was studied comprehensively by King *et al*. (1966). Typical smooth endoplasmic reticulum was found in the lateral branches, corresponding to peri- tracheal glands, which the researchers suggest as being involved in steroid hormone production.

Embryogenesis and phylogenesis

The embryogenesis of the rgl was examined in detail by Poulson (1954a). Amongst other things, the ectodermal character of the lateral, large gland cells, which correspond to the peritracheal glands of Possompes, was shown. There is not much information regarding the phylogenesis of the rgl. Possompes distinguishes four different types of development of the retrocerebral endocrine system in Diptera : (1) *Tipula* type : 2 ca, 2 cc (no equivalent of the pgl has so far been found here), (2) *Chironomius* type : 2 ca, 2cc, and 2 peritracheal glands, each of which are connected with the cc and the ca. (3)Tabanus type: I ca, originationg by fusion of the two glands, 2 cc and 1 peritracheal gland (formed by fusion of two original ones). (4) The true rgl of *Calliphora* with 1 ca (fusion of two), 1 cc and 2 peritracheal glands.

The direct effects of MH

The initiation of the moulting process

The implantation of active pgl on their equivalent results into moulting, not only where secretory activity is naturally suppressed, as in diapause, or where the glands have been experimentally removed, but also in cases where neither the pgl nor the moulting process normally occur, as in adult insects. The moulting process and consequent growth can be prosuced by active pgl along, without a simultaneous implantation of the source of the AH. While, the implantation of an active brain or cc along is not able to induce moulting in the absence of an effective concentration of MH. The resuf was fully confirmed by experiments with isolated MH prepared in crystal form by Butenandt and Karlson (1954).

The circulation of labelled ecdysone injected into the insect body was analyzed by Karlson and Sekeris (1963). They discovered that soon after injection the hormone accumulated in the epidermis and later in the fat body. The nuclei of the epidermal cells exhibited the greatest ratio activity. The authors realized that the sites of action of MH are the nuclei of the epidermal cells and that microsomes of the fat body eliminate the excess of the hormone from the haemolymph.

The regulation of growth

The most important requirement for postembryonic growth in insects, as in any other arthorpod, is the loosening and subsequent freeing of the body from the inexpansible cuticle. MH is therefore indispens- able, though indirectly so, for growth. It is not yet certain whether there is also a direct effect upon growth, as, for instance, the influence of Mh on the mitotic activity which precedes the detachment of the old cuticle. While, there is an increasing amount of evidence to the contrary : *(a)* Growth is possible without any moulting process and thus without MH (e.g., the growth of ovaries and other internal organs as well as of regenerating tissues). *(b)* The implant- ation of agl as well as the injection of ecdysone induces moulting, often without any effect on growth or even with a negative one). *(c)* As distinct from JH, MH alone is not able to induce mitotic activity in larval parts of the body. Without further evidence the presence of MH is thus to be accepted as one of the conditions of growth but there is no cause for accepting it as a true growth hormone of the JH type.

The discovery, that the deposition of endocuticle and the secretion of wax also took place during the intermoulting period in the absence of any MH, led to the conclusion that these processes are not con- trolled solely by MH. MH action after the attain- ment of an active concentration of the hormone in the haemolymph is perhaps responsible only for a change in chitinase activity (or concentration), resulting in the change from chitin synthesis to its decomposition, casuing to moulting and subsequent gradual dissolution of the old endocuticle.

Investigations by Morohoshi and Iijima (1969) were concerned with the effects of the injection of phytoecdysoids (inokosterone, ecdysterone, cyasterone) into starved, ligated last instar *Bombyx mori* larvae and into pupae. In some cases supernumerary larval instars (a VIth instar) resulted. The researchers confirmed that the MHd were concerned ony with the control of moulting and that the structure of the moulted insects depended on JH action. In another study,

Morohoshi *et al.* (1972) injected Vth instar larvae of the same species with varying amounts of ecdysterone at different intervals after the preceding moult. They investigated the effect on the length of the intermoult period

Scrutiny of adaptive changes in larval behaviour prior to pupariation demonstrated interesting correlations with MH action. During 30 hours under dry conditions after the larvae haw left the nutrient medium, they become increasingly sensitive to injected ecdysone. If the dry period is interrupted by a further exposure to moisture, however, their sensitivity diminishes. The researchers stated that this observation might be due to a 'convert' effect of ecdysone. It, seems practical enough however, to assume that leaving the moist environment induces MH production (undoubtedly by activating AH release) and as more endogenous MH is produced, less exogenous MH is needed to induce pupariation. Replacement in the wet medium inhibits AH production, and with it the amount of MH produced up to a given moment, so that more has to be injected to induce puparium formation. The researchers concluded that this was an significant form of phylogenetic adaptation to the given conditions.

In another research paper, Zdarek and Fraenkel (1972) explained that injection of larvae with a large dose of ecdysone under wet conditions shortened the pupariation time to 12 hours, as against 30 hours in the controls. Tanning occurred at various stages of puparium formation (longer, less contracted body, and less smooth surface), again demonstrating the independence from MH and the moulting process of the individual processes of puparium formation. Earlier experi- ments with the facefly *(Musca autumnalis)*, in which the puparium undergoes calcification instead of tanning, proved that this process was equally independent of MH. MH is, of course, the essential releasing factor in this series of processes.

The evolution of typical puparia following the injection of ecdysone into *Calliphora erythrocephala* larvae deprived of their rgl confirms that MH controls contraction of the puparium as well as its tanning and other associated processes. During other set of experiments, Fourche (1969) starved *Drosophila melanogaster* larvae from the 48th hour of the last instar and then supplied them with ecdysone in pure gelation. If the concentration and time of action were adequate, the larvae formed puparia at the proper time. The relationship between endogenous MH and exogenous ecdysone and their correlation with pupariation were discussed. The above results prove that, as in morpho- genesis, these processes are only indirectly dependent on MH.

The action on larval and pupal diapause

But it is quite clear that MH acts directly to break both larval and pupal diapause. Here, however, it must be noted that it is the absence of MH consequent upon the absence of AH which is the immediate internal cause of diapause. A state identical with diapause can be induced experimentally by removal of the source of AH.

The effects on the puffing patterns

The influence on the puffing patterns of the giant polytene chromo- somes in the salivary gland cells and other tissues of *Chironomus* and *Drosophila* will be discussed later .

The action of MH

The impact of MH on the amounts of RNA and DNA during metamorphosis in *Calliphora erythrocephala* was studied by Berreur (1965a) in whole animals and in their separate organs and tissues. The resultant changes, which were tisuse specific for different parts of the body, were not discussed with reference to whether the action of the hormone is direct or indirect. In another investigation, using autoradiography of imaginal discs of the same species, Berreur (1965b) demonstrated that no DNA was synthesized in the absence of MH caused by removal of the ring gland, and that RNA synthesis was limited. Reintroduction of the hormone resulted in nuclear RNA synthesis and, somewhat later, in DNA replication.

Another group of researchers, Sekeris *et al.* (1965) demonstrated that the injection of ecdysone raises the incorporation of radioactive orthophosphate ($H^{32}PO_4^2$) into nuclear ribosomal and messenger RNA in the epidermal cells of *Calliphora erythrocephala*. They accepted that MH stimulated DNA-dependent RNA synthesis. Gillot and Daillie (1968) found a break in DNA production in the silk glands of the *Bombyx mori* at the beginning of each moulting period. They observed asynchrony of the DNA cycle and differences between the two parts of the silk glands, but the maximum amount of DNA per nucleus in the two parts at the beginning of the Vth instar was identical. Neufeld *et al.* (1968) analyzed the effects of injected ecdysterone on *Calliphora stygia* larvae and the conditions for its action on protein and RNA synthesis. Gorell and Gilbert (1969, 1971) demonstrated a similar positive effect by ecdysterone on RNA and protein synthesis in the hepatopancreas of the crayfish *Orconectes viridis*. The particularity of all these effects is yet to be throughly examined.

In *Philosamia cynthia* and *Hyalophora cecropia* a nuclear RNA synthesis in the pgl was activated 3 hours after injecting ecdysterone, whereas similar activity did not start in the ca until 6 hours after administration of the hormone. The administration of MH to pupae which had just began adult development culminated in a drastic decrease in nuclear RNA synthesis.

The action of enzyme activity

Various researchers have expressed that MH interferes with the reaction chain of some of the principal enzymes of the insect organism that regulate normal metabolism and growth, as for instance the cytochrome oxidase system. However interesting and important these findings may be in the study of the corresponding enzyme activity – showing amongst other things the functional importance of insect hormones in research on any question of insect physiology and physiology in general – they do not bring any definite proof in favour of the assumption of a direct participation of MH in the corresponding reactions.

Marmaras *et al.* (1966) discovered an inverse correlation between 5-hydroxytryptophanase activity and the ecdysone titre during the pupal instar of *Calliphora erythrocephala*. Kobayashi and Kimura (1967) illustrated that labelled glucose was converted to haemolymph trehalose in decerebrated *Bombyx mori* pupae injected with ecdysone, but that in the controls (injected with water) it was converted to glycogen. Adenyl cyclase in the pupal wing epidermis of *Hyalophora gloveri* is stimulated after the injection of ecdysterone.

Injected ecdysone has a positive effect on different types of metabolism in ligated last instar larvae of *Musca domestica*. Time and tissue specific incorporation of tritiated thymidine was observed after the injection of ecdysone, but it was limited; however, tritiated uridine was incorporated intensively under the same conditions. Tritiated leucine had a different incorporation pattern. The incorporation of labelled glucose was characterized, in various tissues, by a large increase 1 hour after the administration of ecdysone; it then fell rapidly, at 6-hour intervals until, 18 hours after injection, it reached zero.

The effects on metabolism

Bernardini and Laudini (1966) examined oxygen consumption by the prothoracic glands of young and last instar larvae of two species of cockroach (*Leucophaea maderae* and *Periplaneta americana*) and found that it has cyclical variations. Ultimate oxygen consumption was reached on day before, and minimum consumption one day after each ecdysis. The researchers discussed the possible significance of these changes. Fourche (1967) observed a typical U-shaped curve in *Drosophila melanogaster* pupae, with increases in respiratory metabolism accompanying pupariation, apolysis and ecdysis. In *Thermobia domestica*, Rhdendorf and Slama (1966) found specific oxygen consumption cycles associated with moulting and egg-laying, which corresponded with those in pterygote insects. Oxygen consumption increased during actual ecdysis and fell about half way through the intermoult period.

Many researchers have examined the uptake, transport and utilization of cholesterol and other steroid derivatives by insects. Bergmann and Levinson (1966) tested 15 cholesterol derivatives and isomers in *Musca vicina* larvae and some of them in *Dermestes maculatus* larvae. Both the requirements and utilization of steroids in these insects was discussed. Chino and Gilbert (1971) fed silkworm (*Philosamia cynthia*) larvae on an artificial diet containing ^{14}C-cholesterol. Practi- cally all the substance found in the haemolymph was associated with two lipoproteins. No such incorporation was observed when cholesterol, was added to haemolymph *in vitro*. It was finally accepted that the midgut is necessary for the incorporation of ^{14}C-cholesterol.

A qualitative and quantitative study of haemolymph proteins in *Dixippus morosus* demonstrated a remarkble relationship between them and the moulting process (or morphogenesis), and to yolk formation in adult females. In the final instar larvae, the haemolymph protein concentration reached the maximum on the 13th day after the preceding ecdysis, i.e. at about the time of the new adult apolysis. After that the protein concentration declined abruptly, largely owing to the decline of two glycoproteins, both of which practically disappeared during yolk formation in the adult females.

The effects on rhythmical activity of muscles

Roussell (1970, 1971) explained specific acceleration of the cardiac rhythm in locusts following the implantation of vgl. In last instar nymphs, the implantation of active glands accelerated the cardiac rhythm only in the presence of the ca. In allatectomized adults the implantation of four vgl also acclerated the heart rate. While extirpation of the vgl at the beginning of the last larval instar stabilized the cardiac rhythm at a relatively high level compared with normal specimens. But even here the direct nature of this effect still requires verification.

The indirect effects of MH

The effect on differentiation

Including the effects of MH on growth, many researchers consider that it affects differentiation. While the question of the direct nature of an MH effect on growth may still be considered as a matter for debate, there is no longer any cause for assuming a direct influence of MH on differentiation. However, the dependence of any surface morphological change on the moulting process is unquestion- able. The slightest morphological changes on the surface of the body can only be realized during the relatively short interval between the detachment of the old cuticle and the deposition of the new.

Anybreak in the process of moulting therefore necessarily means the inhibi- tion of growth and differentiation. But the initiation of the moulting process does not in any way mean a simultaneous initiation of differentiation. On the contrary, a premature introduction of MH, as in the ecdysone experiments, results not in more differentiation but in less. This was shown in Fukuda's experiments (1944) with prothoracic gland implantation, and in the experiments of Wigglesworth (1940a, b, 1948) with the parabiosis of moulting specimens of *Rhodnius* combines to those which had passed the critical period for the moulting hormone.

The same effect implies that the result necessarily results from the effects of juvenile hormone on growth and differentiation. By implanting active ca the gradient growth, resulting in differentiation, may be changed into an isometric one without change in form. Such an inhibition of differentiation does not mean any inhibition of the moulting process which is distinctly accelerated. There is no basis for assuming any antagonism between MH and JH. Differentiation can be completely separated from both moulting and growth and there is no connection between the effects of Mh on one or the other. It is confirmed that the influence of MH on differentiation is therefore purely indirect.

Joly (1958) when analyzed the effects of the ventral gland hormone on the mitotic activity of the epidermal cells and an allometric growth in *Locusta* during metamorphosis, and found that MH only affects the moulting process.

The postulation that MH has only an indirect effect on growth and protein synthesis is supported by the following experiments. Chilled *Antherea polyphemus* about to start adult development and ligated abdomens of *Hyalophora cecropia* and *Samia synthia* adults were injected with crystalline ecdy- sone together with mitomycin C (which inhibits DNA replication). Ecdysone was found to cause apolysis in the pupae of these insects, in the presence of mitomycin, but growth, morphogenesis and the rest of the moulting process were inhibited. If ecdysone were administered after DNA synthesis, the epidermal cells also responded by depositing new cuticle. Similary process took place in adult nsects, in which no more DNA synthesis takes place, but the entire moulting process can be activated experimentally with the same ecdysone.

Courgeon (1969, 1970) analyzed the effect of MH on the imaginal discs of eyes and antennae in tissue cultures. Generally mitotic activity stopped a few hours after explanation, but if an active ring gland was implanted at the same time it continued for up to five days. When the nervous system (the cerebral hemispheres and the ventral ganglion) was also added, there was a stronger effect.

The impact of MH on morphogenesis were examined in detail in ligated *Galleria mellonella* larvae. Repeated dipping of isolated abdomens in methanol solutions of ecdysterone in different concentrations resulted in morphogenetic changes of varying degrees after ecdysis. Different types of cuticle were obtained in different changeing from a larval type, via a more or less normal pupal type, to an imaginal type with more or less developed scales. Taking into consideration the amount of hormone applied (the concentration and number of dippings) and the interval between treatment and subsequent moulting, the following explanation is submitted, in agreement with the gradient-factor theory. The degree of morpho- genesis depends upon the length of the interecdysial period, or, to be precise, upon the time between apolysis and deposition of the new cuticle. If this is shorter than normal (after a large dose of the hormone), the degree of differentiation is smaller and the cuticle retains a more or less larval of character. If the MH concentration is low, the interecdysial interval is longer and morphogenesis can progress further. This is the condition with omission of the pupal type of cuticle. Moulting is not caused by small does, but certain developmental changes in the internal organs do occur .

The action of ovarian development

The indirect character of the morphogenetical impact of MH is quite clear from its effect on the development of the ovaries. It was demonstrated by Strich-Halbwachs that the removal of the vgl at the beginning of the last larval instar in *Locusta migratoria*, which prevents moulting, also suppresses the development of ovaries but to a limited extent. The oocytes, which average 0.7 mm in length at the time of the operation, reached 1.8 mm within one month in the permanent larvae thus obtained, while they reached 7.2 mm in the metamorphosed controls at the moment of egg-laying. Few oocytes in the experimental series were abnormal or had even degenerated. It is evident that the same factor is affecting here, i.e. the inhibitory influence of the inextensible cuticle, the only difference being that its effect on the internal organs is less direct than on the epidermis. The relationship between the moulting process and the development of the ovaries also permits a rational explanation of the findings of Bodenstein (1953a-d) and Wigglesworth (1957) in which the induced nature of this type of MH effect is also evident.

The effects on ovarian development. The injection of ponasterone A affects the ovarian development (the weight of the ovaries and the number of eggs) in isolated pupal abdomens of *Bombyx mori*. It also had other morphogenetic effects. Ovarian weight and the number of eggs increased pro- portionally to the dosage up to an amount of 2.5 μg per specimen. No change was observed after larger doses, which resulted in the formation of a scaleless adult cuticle. The implantation of vgl into adult *Locusta migratoria* females, stimulated growth of the oocytes, but caused the more developed oocytes to degenerate. The activity of the accessory glands and the genital apparatus was stimulated in allatectomized locusts of both sexes.

One may even include that the independence of the development of the ovaries from the moulting process and MH is quite evident from the fact that it normally takes place in adults where no prothoracic glands exist and no moulting occurs.

The action on the corpora allata and reproduction

The entomologist Engelmann (1959) examined the effect on the function of the ca and the development of ovaries of implanting active larval pgl into adult females of the cockroach *Leucophaea maderae*. He obtained moulting of the adult in some cases and nearly always a partial inhibition of ca secretion. He describes this as a direct effect on the ca and an indirect one on the neurosecretory brain cells. Both these inhibitory effects were avoided by the simultaneous implantation of a brain and suboesophageal ganglion. An injection of ecdysone had an effect similar to that of the implanted prothoracic glands. In both cases cell division was found in the ca with a subsequent increase in the number of cells. As suggested by, Engelmann, the results for a related species, *Periplaneta americana* to some extent contradict those obtained by Bodenstein (1953a, e). There is also a remarkable discrepancy in the conclusions of both these researchers since the pro- thoracic gland secretion is produced during larval instars almost simultaneously with that of the ca without producing the slightest inhibitory effect. Some results in recent years have even shown that the ca may cause a direct positive effect on the induction of moult- ing which has no connection with JH. Nevertheless, there is the possibility of a single simple explanation of all these facts provided in the *'law of correlation in consumption'* by the different tissues inside the body. When comparing the findings of Engelmann with those of Bodenstein, two major points emerge. Firstly that the moulting process, even when not accomplished, necessarily consumes part of the available body reserves. This deficiency becomes clear in the reduced activity of the corpora allata and in the restricted growth of the ovaries. Secondly, such a serious interference with the physiological processes of the body as that of allatectomy and the resulting regeneration processes may in itself produce an impulse strong enough to prolong the activity of the pg.

In *Locusta migratoria*, the disconnection of ncc and the resulting increase in the volume of the ca has no inhibitory effect on the moulting process. In this context it is interesting to note the effect of folic acid on diapausing pupae of *Pieris brassicae*, where it increases the amounts of ribonucleic and desoxyribonucleic acids causing tumour growth in the absence of MH . The researcher attempts to explain these findings on the priniciple of artificial induction of endogenous viruses.

The action on regeneration

Other aspect of differentiation is also true for regeneration : that is, it is also independent of moulting. However, there is thought to be an indirect interdependence, at least as far as more extensive regeneration is concerned. It can either be acclerated by the action of MH, or suppressed or greatly delayed by the removal of MH. Less extensive regeneration, including mitotic activity, may occur independently of MH.

A study of contrasting features of the activity of the pericardial glands in normal and regenerating (with the mesothoracic legs removed) Ist instar nymphs of *Carausius morosus* was made by Vietinghoff, Penzlin and Spannhof (1964). A statistically significant difference in the volumes of nuclei was found between the glands of regenerating and normal specimens. The decrease in volume of the nuclei in normal specimens connected with the onset of the moulting process, was delayed by two days in the regenerating individuals.

Many researchers have examined the effect of injury and wound healing on local cuticle formation on the general moulting process (Krishnakumaran, 1972). In both cases the effect was positive. The effect of injury on the general moulting process seems to be mediated by the brain, since brains from injured specimens, unlike those from the intact controls, displayed AH activity on both the pgl and ca. In the case of local cuticle formation, which was also observed in adults, the presence of MH does not seem to be absolutely necessary for the cuticulogenic activity of the epidermal cells. This is also pointedout by the findings of Gersch (1972), who observed chitin hypertrophy in the salivary glands of *Periplaneta americana* following injury in the thoracic region (pgl extirpation).

Schaller and Andries (1970a, b) and Schaller (1972a, b) analyzed the regeneration cells in *Aeschna cyanea* (Odonata) larvae deprived experimentally of their vgl. Mitotic activity in these larvae was rather low, but the implantation of active vgl resulted in an abrupt and significant increase in the activity of these cells. In another research paper, Schaller (1970) concluded that MH was responsible both for the activation of remaining regeneration cells and for regeneration of the midgut epithelium in the intermoult period, although these cells, like the cells of the eye imaginal discs, also displayed lower mitotic activity in the absence of the hormone. On the other hand, no mitosis was observed in the absence of MH in permanent larvae. A detailed study of the eyes and optic lobes in permanent larvae of the same species fully confirmed the above conclusion. Keeping in consideration the gradient-factor theory, it emerges as an interesting observation.

The influence on the morphological colour change

Remarkably visible morphlogical colour changes occur in various insect species during metamorphosis. Few exhibit this in the adult stage, but in the majority it is a transient phenomenon. These changes may be quite affected by the suppression of premature introduction of MH and can either be suppressed or prematurely induced. Because of this, many researchers have assumed that the moulting hormone exerts a direct effect on the chemical synthesis of the corresponding pigments. Buckmann (1956-1959), for example, in his extensive studies of morphological colour changes in *Cerura vinula* larvae during metamorphosis reached the conclusion that these changes are caused by low concentration of the MH, whilst large doses of the same hormone induce pupation. Generally the green caterpillars turn dark at the time the cocoon is made due to the production of a red ommochrome, which first appears in the epi- dermis and one day later in the fat body. The morphological process of the pupal moult does not start until five days later. The colour changes are completely halted by ligaturing the part of the body lacking the MH. When 66

Calliphora units were injected into such a permanently green abdomen, they cause reddening of the epidermis. Three hundred and thirty *Calliphora*-units caused reddening of the fat body only, whilst very large doses, 330 to 6600 units caused pupal moulting without any colour change. Buckmann (1947a) concluded that there the effect of ecdysone could be due to a simple reduction of the ommochrome pigment and he suggested that 'das ware ein idealer Modelfall einer chemisch einfach definierten Hormonwirkung'.

Even though an alternate, direct dependence of a pigment reaction on the hormone cannot be excluded, the following explanation seems to be the most probable : the effect of MH consists, as in other-metamorphosic processes, in the conditioning of species specific chemical and morphological changes, in the normal course of development. The discovery that removal of the ca (by decapitation) in the earlier, larval instars result in similar changes in the larval periodfullly accords with this.

Doubtlessly a similar dependence would be found in all other cases of morphological colour change described for various other insect species, with an experimental approach similr to that used by Buckmann with *Cerura*. It is confirmed from the experiments of Wigglesworth (1940) with *Rhodnius* that there is no direct influence of MH on pigment formation : in the imaginal moulting of this species black pigment vanishes from in some places where it wwas earlier present.

The correlation between the prothoracic (= ventral) glands and phase development

This was examined in *Schistocera gregaria* and *Locusta migratoria*. *The prothoracic glands which gemerally disappear in adults were found to survive for a longer period in the solitary than in the gregarious phase. Besides this the ca are larger in solitary than in gregarious nymphs. Extirpation of the greater part of one of the pair of the prothoracic glands in solitary nymphs resulted in the assumption of the gregarious habit. It is accepted that a complex interaction exists among all three metamorphosis hormones in influenceing phase differences.*

The action on the development of internal parasites

The group A virus *Sindlois arbovirus* is known to reproduce during the metamorphosis of *Calliphora erythrocephala*. In larvae deprived of their MH source by ring gland extirpation, no multiplication of the virus takes place.

The control of MH production

The secretion of the MH by the pgl and their equivalents is influenced by the presence of a certain effective concentration of AH in the hemolymph. It is therefore the control-system of AH which, initially, also controls the production of MH. It was shown by the experiments of Williams (1952) that pgl may be activated to produce MH even in the absence of AH by the introduction of MH. Thus, diapause in *cecropia* pupae may be broken by the implantation of active pgl or by the injection of ecdysone into pupae with inactive pgl, even in an amount which itself would not be sufficient to initiate the moulting process (in specimens with their pgl removed). On the basis of these experiments Williams suggests that co-ordination of secretory activity between the left and right pgl is definite. Further- more, extensive or repeated injuries and the regeneration resulting from these injuries are brought about by the re-activation of pgl. They may even in some conditions cause supernumerary moulting in adult insects. But, such effect is probably restricted to the phylogenetically 'lowest' groups of Hemimetabola such as cock- roaches, not yet very far remote from the Apterygota where moulting persists in adult insects.

New information on this subject is available in recent years. Zdarek and Fraenkel (1971) studied neurosecretory control of MH release during pupariation of *Sarcophaga argyrostoma*. In an earlier paper they demonstrated that there is strict environmental control of the initiation of the pupa- riation process in Cyclorrhapha. The larvae do not pupariate unless they leave their wet feeding medium and spend one or two days in a dry environment. However, implantation of the cerebral and ventral ganglion, together with the rgl, induced pupariation even in wet larvae. If the rgl alone was implanted, only larvae which has spent some time (insufficient by itself) under dry conditions pupariated. Neural connection with the cns proved to be essential for an inde- pendent pupariation effect. The activity of the rgl consistently increased in the pre-pupariation period in active flies.

Gersch and Sturzebecher (1968, 1970,) carried out a detailed examination of AH and its effect on the pgl in *Periplaneta americana*. The pgl were experimentally activated by one larval brain of *Antherea pernyi* of three brains of *Periplaneta americana*. After 8 hours' exposure to AH, a substance was produced in the pgl. Thin-layer chromatography and mass spectrometry have confirmed this susbtance to be a non-separated mixture of stearic, oleic, linoleic, linolenic, palmitic and myristic acid.

According to further discoveries on this subject, some juvenoids were found to activate diapausing or decerebrated *Antherea polyphemus* and *Hyalo- phora cecropia* pupae . Their effects on the pgl were proportional to their juvenilizing activity.

Sehnal and Edwards (1969) examined the inhibition of pgl activity by body constraint in *Galleria mellonella* larvae. Constraint inhibited AH secretion and the authors interpreted this as nervous inhibition (possibly owing to the elimination of an environmental nervous impulse). MH production was also inhibited, although the epidermis remained capable of its moulting function (demonstrated by the administration of ecdysterone). This indicates that an appropriate external or internal stimulus is needed for activation of the neuroendocrine system.

Ohtaki *et al*. (1968) examined the rapid decrease in MH activity in the insect organism after hormone administration. They concluded that a special inactivating mechanism was present in the tissues, but not in the haemolymph. When injected into mature *Sarcophaga peregrina* larvae, 1Mg ecdysone lost 50 per cent of its activity in 1 hour and by 8 hours only 2 per cent remained. In the critical phase for pupariation, the larva contained only 7 per cent of one *Sarcophaga* unit. According to the researchers, this could have been because of accumulation in the tissue of convert effects determining the overt response. No accumulation was observed in the haemolymph. A more practicle explanation, however, is that early changes in the target tissue (the ectodermal cells) are reversible up to a given point of time, until then the presence of MH is indispensable; but not later than that

Chemical characteristics of MH

Isolation of MH

The first prothoracic gland hormone of the insect metamorphosis hormones to be prepared was in crystalline form. Butenandt and Karlson (1954) used 500 kg of silkworm pupae to prepare an extract using the following technique.

The pupae preserved in methanol were subjeced to a pressure of 50 atm and the resulting extract reduced by evaporation *in vacuo*. The 30 litres remaining from about 650 litres of the original extracts were shaken several times with butanol. The separated butanol extract was then washed successively with 1 per cent H_2SO_4, 10 per cent Na_2CO_3 and 5 per cent acetic acid. It was evaporated and the remaining red-brown oily substance (79 g) was redissolved in butanol, and after filtration through about 800 g alluminium oxide, following Brockmann (activity V), re-evaporated and dried *in vacuo*. It was then dissolved in water, shaken successively with ether and ethyl acetate and the water phase again evaporated. The remaining 1.15 g of red-brown oil showed an activity of 1 *Calliphora*-unit in 0.5 g. It was then purified by chromatography in aluminium oxide activity IV and, using ethyl acetate-butanol (3 : 1) and ethyl acetate- methanol (3 : 1), 167 mg of a highly active fraction (1 *Calliphora*-unit in 0.1 μl) was obtained. After recrystallizing from ethyl acetate and methanol, 25 mg of crystalline substance was obtained which depicted an activity of 1 C.u. in 0.0075 Ml. It is called endysone and has been identified with MH.

Chemical characteristics

This crystalline preparation formed silky fibrous needles. These change their appearance at 161-162°C probably because of the loss of water of crystalization, and melt at 235-237°C. Within these limits the substance is heat stable; it is soluble in water and very soluble in methyl alcohol; it shows distinct absorption bands in the ultraviolet and infrared parts of the spectrum. The ultraviolet spectrum reaches its maximum at 244 nm (in ethyl alcohol) and at 249 nm (in water); and the infrared at 612 nm. It is optically active :

$$(d)\frac{20^{\circ}}{D} = +58.5^{\circ} \pm 2^{\circ} \quad (d)\frac{21}{546} = 82.3^{\circ} \pm 2^{\circ} \qquad (2)$$

(at a concentration of 7 mg in 2 ml ethyl alcohol).

Depending on these findings the presence in the molecule of α and β unsaturated keto groups was assumed. The infrared part of the spectrum also suggests the presence of at least two hydroxyl groups. It was further inferred that a $_2CH_5$ group and perhaps also a CH_3 group are present. Both acids and bases easily inactivate the substance. There is another band in the spectrum at 265 nm which may be assumed to have originated by the splitting off of water which led to the formation of the double unsaturated system.

The hormone decomposes quickly in alkaline solutions, the decomposition being accompanied by a successive change of spectrum. The first analysis of the dried substance showed elements in the following ratio : C_4, $_4H_7$, $_3O_1$. The molecular weight was originally shown to be about 300; and the emperical formula $C_{18}H_{31}O_4$. Concentrates from *Crangon vulgaris* which were found to be more active in crustaceans than in insects point to the existence of several different chemical and physical characteristics (e.g. increased polarity) in the crustacean hormone. Further research has confirmed that at least two different substances are present with identical physiological activity, which are separable into α- ecdysone and β-ecdysone (now ecdysone and ecdysterone, by extracting with a cyclohexane-butanol-water mixture. The two susbtances may be separated by their melting points and solubilities though their spectra are very similar.

Karlson and Hoffmeister (1963a) worked incessantly on the chemical properties of ecdysone. 1000 kg of dried pupae of *Bombyx mori* removed from cocoons obtained from the silk factory were used. From these 250 mg of crystalline ecdysone with a specific activity of 100 000 *Calliphora*-units per mg was obtained.

With the help of of X-ray analysis and mass-spectroscopy a molecular weight of 464 was confirmed. This corresponds to the empirical formula $C_{27}H_{44}O_6$. Evidence was obtained of the steroid nature of ecdysone, one oxy-group andfive hydroxyl-groups were found. On the basis of the empirical formula a relationship with cholesterol was suspected. Injection of tritium-labelled cholesterol into *Calliphora* larvae yielded crude radioactive ecdysone after extraction and purification. It was then confirmed that cholesterol is a precursor of ecdysone. A similar observation was made by Kanazawa and Teshima (1971) with C^{14}-labelled cholesterol in the spiny lobster

Panulivus japonicus. The activation of the prothoracic glands in decerebrated *Bombyx* by ether extracts of brains and by cholesterol, found by the Japanese authors, could probably be explained by this. It in no way contradicts the fact that cholesterol is inactive in diapuasing sawfly larvae.

The substance proved to be active not only in various Lepidoptera but also in Hymenoptera, and Hemiptera Karlson (1957) later succeeded in isolating a similar susbtance from the shrimp *Crangon vulgaris*, using the same method. The *Calliphora*-test shows a distinct, but much lower activity in this substance (1 C.u. = 50 mg of the dried oil;. Similar preparations have been obtained from *Carcinus moenas* and *Portunus holsatus*.

Almost similar and yet more active MH preparation has recently been obtained by Stamm (1958) from the grasshopper *Dociostaurus maroccanus*. The researcher used 10 kg of adult specimens which yielded 11 mg ecdysone and 12 mg ecdysterone in crystalline form. The substances were shown to be chemically identical with the silkworm α- and β-ecdysone and were found to be more active in *Calliphora*-tests. The reduced activity of the silkworm preparations is assumed to be due to the presence of kynurenin and other substances harmful to *Calliphora*.

A number of research papers by the Karlson school deals with the question of the chemistry of the MH and the biochemistry of its production. On the basis of their work the definitive solution of the chemical structure of ecdysone was made by Karlson (1965). As stated earlier, it is a steroid molecule of the following structure :

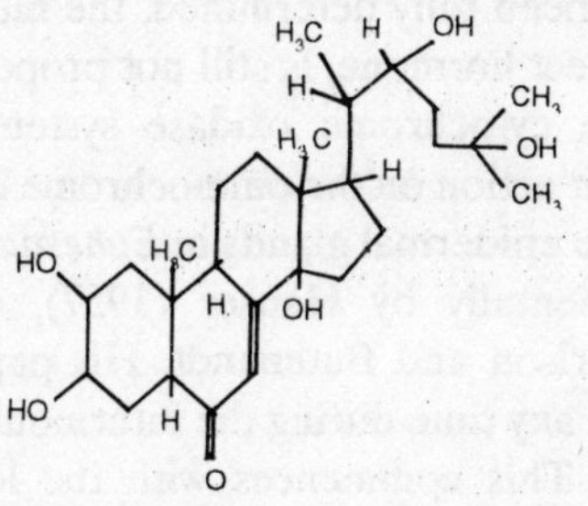

It is a 2β, 3β, 22βF. 25 = pentahydroxy-Δ7-5β-cholesteron- (6). The C_{20} has the normal steric configuration, i.e. 20βF-methyl 6-keto-0 = 0 − 1; 14α − 0 = 0 − 2, 2β − 0 = 0_3, 3β0 = 0 − 4, 25 − 0 = 0 − 5, 22 − 0 = 0 − 6.

Five various biologically active fractions of ecdysone are assumed to have been obtained by chemical treatment of *Bombyx mori* extracts, three of them being undescribed and two known from the work of Karlson and his colleagues and earlier from that of Burdette and Bullock (1963). A positive effect on tissue growth in mammals is claimed together with stimulation of protein synthesis in the cytoplasm.

Immediately after Karlson's original formula for α- ecdysone was discovered, the compound was simultaneously synthesized in laboratories in West Germany and the U.S.A. (Furlenmayer *et al.*, 1966a, b, 1967; Siddall *et al*., 1966a, b) and its structure fully confirmed. Furlenmayer *et al*. started with ergosterin and used 14 steps. The main steps were (20S)-2, 3β- diacetoxy-20-formyl-5α-pregn-7-ene-6-one and (22R)-2β, 3β-diacetoxy-14-22-dihydroxy-25-(tetrahydropyran-2-yloxy-)-5α-cholest- 7-ene-6-one. Later ecdysterone was also synthesized.

As soon as the chemical structure of ecdysone and its artificial synthesis were resolved, the number of studies on its various chemical properties and those of its many natural and synthetic derivatives rapidly multiplied.. The most complete survey is the latest one by Slama *et al*. (1973), which is yet to be published.

The mode of action of MH

Inspite of the fact that the chemical composition of the moulting hormone has now been fully determined, the mode of action of this, the best known insect hormone, is still not properly understood. The way it affects the cytochrome oxidase system has already been mentioned, as has it action on the ommochrome changes. Its effect on development of the epidermal glands in *Ephestia kuehniella* has been examined experimentally by Hanser (1957), using the ecdysone preparation of Karlson and Butenandt. His paper shows : (1) The injection of MH at any time during the intermoult period induces the moulting process. This commences with the loosening of the old cuticle and the secretion of the moulting fluid, and continues with the formation of a new cuticle (pupal cuticle). (2) A single dose, even if rather strong (40 C.u. or more), is inadequate for the completion of the moulting process. If no further dose is supplied the moulting process stops in one of its later stages and can only persist when more ecdysone is injected.

It can be confirmed from the findings of Hanser (1957) and the results of earlier experiments with parabloses, and pgl transplantations that MH influences the activation of all the enzyme of the epidermal cells. The first phase of this activation consists of the swelling of the cells, which may or may not result in mitotic activity. It is followed by the second phase, the secretory activity of the cells, resulting in the production of moulting fluid, the initial effect of which is the loosening of the old cuticle by chitinase which subsequently dissolves the inner layers of the cuticle. In the third phase, the new cuticle is deposited and the secretory activity of the cells continues yo produce the first layers of the new cuticle, most probably under the action of the same enzyme (chitinase). The phases generally overlap one another. It seems that the presence of MH is imperative at least to begin all these phases.

The consequenece of thiourea injections on moulting and pupation in silkworm *(Bombyx mori)* larvae were exmined by Chmurzynska and Wojtczak (1963). Injections up to 24 hours before larval moulting or cocoon spinning resulted in the retention of the old cuticle and accumulation of a haemolymph-like fluid between the old and the new cuticles. Partial inhibition and a dealy in the sclerotization of pupal cuticle was observed in histological sections.

Advancement in the technique for isolating MH in small quantities have made it possible to compare the MH titre at various stages of an instar and in various periods of postembryonic development (Shaaya and Karlson, 1964; Shaaya and Sekeris, 1965; in *Calliphora erythrocephala*). There are two maxima – one just prior to pupation and the other on the 3rd to 5th day of the pupal instar. The first peak correlated with puparium formation, and the second with histogenetic processes. No ecdysone could be detected on the last two days of the pupal instar. The same method was employed to examine the activity of several enzymes was studied in a similar manner.

Many other research investigations were carried out to examine , the ecdysone titre in relation to body weight, in two lepidopteran species – *Bombyx mori* and *Cerura vinula*. In *Bombyx mori* a maximum was found in the IVth (penultimate) and Vth larval instar shortly before ecdysis; immediately before ecdysis occurred MH fell to zero. Two maxima were observed in the pupal instar. Some ecdysone was still present after adult ecdysis (about four times more in the female than in the male), but by the 8th day it had virtually disappeared. Two maxima were found in *Cerura vinula* – one at the beginning of the colour change and the other shortly before pupation. No ecdysone was detected in the pupal instar. But, the information may not corre;ate to the amount of active MH in the haemolymph. Adelung (1966) estimated MH titres similarly in the crab *Carcinus moenas*, and discussed their variation in the context of the crustacean endocrine system, i.e. the X organ-sinus gland complex and the rostral gland (Y organ).

The number of individual secretory and morphogenetic processes which take place in the epidermis during moulting cycle and their a hormonal dependencies were discussed by Locke (1965). From the results of experiments on the moth *Calpodes ethlius* (Hesperidae), Locke asserted that whereas the moulting process is controlled only by ecdysone, intermoult events such as wax secretion and endocuticle deposition need the assistance of a neuroendocrine factor from the brain.

A combination of recent views and findings on moulting control was published by Weir (1970), who on the basis of his experiments on *Calpodes ethlius* classified three phases of development in the last (Vth) larval instar. The first is the growth phase, from ecdysis up to 66 hours. After that the brain is no longer required for the induction of pupation, and the next phase, up to 156 hours, ends when the pgl are not longer necessary. Pupation takes place at 192 hours. From an analysis of the data on the pgl, the researrcher concluded that certain abdominal tissues itneracted with the pgl in the induction of the moulting process and stated that the oenocytes might be involved.

THE JUVENILE HORMONE (NEOTENIN) (JH)

(= inhibitory hormone, Wigglesworth, 1935; = status quo hormone, Williams, 1952; = das Larvalhormon, Weber, 1954; = CAH = corpus allatum hormone, Slama, 195; = 'soldier hormone', Luscher, 192; = gonadotropic hormone, Engelmann, 1957)

The metamorphosis-preventing activity of the corpus allatum was initially demonstrated by Wigglesworth in his classic parabiosis experiments with the blood-sucking bug *Rhodnius prolixus* in 1935. Wigglesworth (1936) demonstrated that joining a nymph in the penultimate or an earlier instar with a nymph at the beginning of the last larval instar results in the partial or total suppression of metamorphosis without preventing an increase in size. In this way VIth instar giant nymphs imagos. Conversely, when the ca of the younger nymph was removed by decapitation, a miniature precocious imago resulted.

Wigglesworth not only the source of this inhibitory effect but was also the first to provide definite evidence of its hormonal nature by demonstrated that the same results are produced when the specimens are connected by a glass tube so that only the haemolymph communicates. The original term 'inhibitory hormone' was soon changed to 'juvenile hormone' after the active role of the hormone in producing larval characters in adults was recognized. This discovery was confirmed by a number of other authors for various insect species, Bounhiol (1936, 1938a) for *Bombyx mori*; Pflugfelder (1937a, b) for *Carausius morosus*; Piepho (1938b-d) for *Galleria mellonella*; Bodenstein (1938b) and Weed-Pfeiffer (1938) for *Melanoplus differentialis*.

Contemporary researcher, Wigglesworth (1935, 1936) also discovered the effect of the ca hormone on ovarian development. This was later confirmed for a number of other species whilst in others, such as *Bombyx mori*, no ovarian control by a corresponding humoral factor was observed. The other JH effects were ascertained subsequently. The apparent contradiction between the inhibitory, metamorphosis- suppressing effect of the ca hormone and its total metabolism- increasing influence, suggested to a number of researchers that there were two or more ca hormones having different effects. However, this hypothesis theory was not subsequent experiments, and it was soon recognized that all the available data could be explained by the operation of a single substance, JH, possessing the character of a growth hormone. Since then experimental observational evidence has accumulated in favour of this explanation. Interordinal non-specificity of JH was shown by the transplantation of ca between Blattoptera and Hemiptera, and Phasmida and Lepidoptera. The first attempts at elucidating the chemical nature of JH were made by Williams (1956), Schneiderman and Gilbert (1958, cf. Schneiderman, 1960), using ether extracts of male abdomens of the Saturniid *P. cecropia*. Its chemical structure was first determined by Roller and his co-workers (1965a, b, etc.). Many other chemically different substance, i.e. the juvenoids behaved in the same manner.

The corpora allata (ca)

The first description of the insect ca is probably that of Muller (1828), in *Blatta orientalis*; however, he expected the glands to be nervous in character, believing visceral ganglia supply the dorsal aorta and the anterior portion of the gut. This analysis was accepted by a number of other researchers. The first doubts regarding the nervous character of the ca were expressed in 1899 by Janet, from his detailed anatomical study of the female of the ant *Lasius niger* and to whom we owe the term corpora allata. In the same year, Heymons (1899) also concluded that the ca in *Carausius morosus* were non- nervous. The first systematic, comparative morphological study of the insect ca was carried out by Nabert (1913), who clearly expressed the opinion that they were endocrine in function. Since their effects were recognized by Wigglesworth (1935) the number of research papers dealing with their structure and function has been published in great number.

Morphology

The ca are found in all groups of winged insects in which they have been searched for , and in some Apterygota, where they are very primitive. Several different types of ca can be distinguished, on the basis of structure, position, relationships with other endocrine glands and phylogenetic considerations. The simplest and most primitive type known are the bladder-shaped ca of Thysanura and stick-insects, which in some species still keep their original position on the lateral epidermis. They are followed by the paired lateral ca with cc, as in Ephemeroptera, Odonata, etc..

The direction of development at this stage is through the approximation and later fusion of the ecc, as in Blattoptera, Orthoptera, Mecoptera and some of the Diptera, e.g. *Culicidae*. The other tendency was to fuse the ca which can take place either below the dorsal aorta as in Hemiptera, or, above it, as in Tabanidae. A special type with both cc and ca fused and connected by the prothoracic glands fused with the nervi allati to form a single structure, is represented by the ring gland of Diptera Cyclorrhapha.

Cyclical changes occur in the ca in each larval instar as well as in the adult. The ca in the plant-feeding bug *Oncopeltus fasciatus* have their smallest volume at the beginning of each larval instar and increase gradually in size to reach a maximum at about the middle of the intermoult period. Then they reduce in size with an associated decrease in the secretory activity before the next moult, the minimum size at the beginning of each instar being distinctly less than the maximum in the earlier one. The increase in size of the ca in each larval instar thus has two components – an increase due to secretory activity (i.e. the difference between the maximum volume during an instar and that the next ecdysis), and growth of the gland itself (i.e. the difference between its volume at one ecdysis compared with the previous one). In the last larval instar a distinct increase in the volume of the gland also occurs, but is compartively less than in the earlier instars. In the adult stage a clear increase in the size of the gland byins 3 to 4 days after imaginal moulting and continues for the whole period of egg-laying (from the 8th and 30th day), when the volume of the ca increases by about 30 times compared with that immediately after moulting. Legay (1950) found a distinct increase in the volume of the gland in *Bombyx* at the time of each larval ecdysis and a very small

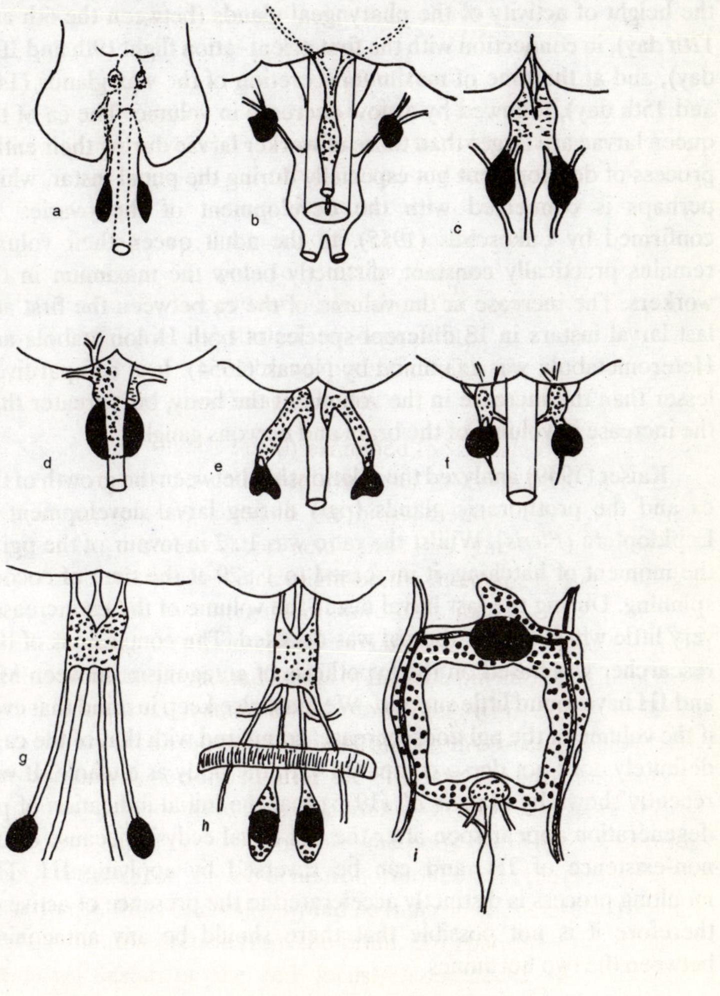

increase during the intermoult periods. This type of growth is probably connected with the displacement of the critical period for AH and thus, necessarily, also of that for JH towards the end of the intermoult period. It is a secondary adaptation in the phylogenetically 'highest' insects, connected with the shortening of the moulting process. The development of the ca in *Carausius morosus* and in both larvae and adults of the honey bee *(Apis mellifica)* was systematically examined by Pfulgfelder (1958). He found a remarkable increase in the volume of the ca of worker bees in the first days of imaginal life, at the height of activity of the pharyngeal glands (between the 6th and 12th day), in connection with the first orient- ation flight (9th and 10th day), and at the time of maximum secretion of the wax glands (14th and 15th day), followed by a slow decrease in volume. The ca of the queen larvae are larger than those of worker larvae during their entire process of development but especially during the pupal instar, which perhaps is connected with the development of the ovaries, as confirmed by Lukoschus (1955). In the adult queen their volume remains practically constant, distinctly below the maximum in the workers. The increase in the volume of the ca between the first and last larval instars in 18 different species of both Holometabola and Heterometabola was examined by Novak (1954). It is comparitively lesser than the increase in the volume of the body, but greater than the increase in volume of the brain and nervous ganglia.

Kaiser (1949) analyzed the relationship between the growth of the ca and the prothoracic glands (pgl) during larval development in Lepidoptera *(Pieris)*. Whilst the ratio was 1 : 2 in favour of the pgl at the moment of hatching, it increased to 1 : 29 at the time of cocoon spinning. During the last larval instar the volume of the ca increased very little whilst that of the pgl was doubled. The conclusions of this researcher was based on the hypothesis of antagonism between MH and JH have found little support. We must also keep in mind that even if the volume of the pgl does increase, compared with that of the ca, it definitely does not do so compared with the body as a whole. It was recently shown by Mala *et al.* (1975) that the initial indication of pgl degeneration appear soon after the last larval ecdysis because of the non-existence of JH, and can be reversed by applying JH. The moulting process is distinctly accelerated in the presence of active ca therefore it is not possible that there should be any antagonism between the two hormones.

In *Hyalophora cecropia* adults, an extraordinary sexual dimorphism in the size of the ca was discovered by Schneiderman and Gilbert (1959) in contrast to that in bugs and most other insects. Whereas the pupal ca show the same weight of about 30 μg in each sex, those of adult males increase to about 400 μg whereas to those of the female adults only increase to about 65 μg. This is in agreement with the JH activity of ether extracts of male and female abdomens, that of the female beig only about 1/40th that of the male. A similar variation is observed in the ether extracted liquids (0.201 g in the male and 0.045 g in the female), which is correlated with the much greater flight activity of themale. Similarly, Rehm (1951) found a 20-fold increase in the ca adult male *Galleria mellonella*, whereas the increase in the female was only 3-fold. But the number of cells per ca is, almost equal in each sex.

Histology

The ca are sphericae, ovoid or pad-shaped bodies in most insects, connected with the cc by a nerve (nervus allatus) which may become reduced so that the two kinds of glands are closely approximated or, in some cases, partially fused. They are enclosed in a mesenchymatous connective tissue membrane formed by plate-like cells with an oblong nucleus and homogeneous surface layer. Mendes (1948) found the following three types of glandular cells in the ca of *Melanoplus* – undifferentiated cells, normal secretory cells and polyploid giant cells. A remarkable feature of the glandular cells of the ca is that they increase in size considerably during the peirod of their secretory activity and shrink during their inactive periods. The period with histological signs of secretory activity coincides with that of their hormonal activity as shown by parabiosis and transplantation experiments. There is no histological proof that more than one ca hormone is produced in any of the species studied.

The ca can be classified into four groups, according to their histological structure :

(i) *The epithelial vesicular type*, consists, in a central cavity, secretion *(as in Aeschna)*, connective tissue *(as in japyx)* of a system of concentric lamellae *(as in Phasmidae)*. This is the original, phylogenetically 'lowest' type (even though secondarily simplified in some cases), showing the ectodermal origin of the ca.

(ii) *The pseudolymphoid type*, structured by small cells with reduced deeply staining cytoplasm resembling that of lymphoid tissue in verte- brates. Ca of this type have been discovered in Ephekerida and Odonata.

(iii) *The small cell type*, which is the most common, is found in Blattoptera, Orthoptera, Hemiptera, Coleoptera, etc. It is structured by numerous small irregularly shaped cells, in the intercellular lacunae of which, secretion accumulates during periods of activity. The nuclei are egg-shaped to oblong, their cytoplasm being relatively clear.

(iv) *The large cell type*, phylogenetically the most evolved, and it is found in Panorpata, Hymenoptera, Trichoptera, Lepidoptera and many Diptera. The gland consists of few large cells with large, often lobular nuclei and numerous uniformly distributed chromatin granules.

The histophysiological changes in the ca in adult *Leucophaea maderae* were examined compreyhensively by B. Scharrer and Harnack (1958) and Harnack (1958a, b). No great changes in the low secretory activity of the male ca were observed during the whole adult life except for a short 'activation period' just after imaginal moult. While in females significant changes occur in connection with the ovarian cycles. A conspicuous increase in the amount of cytoplasm, both absolute and relative (in relation to the size and number of nuclei), occurs during the activation period. The researchers describe the changes minutely. Harnack (1958a, b) also observed the effect of starvation on imaginal moulting. The ca volume decreased markedly shortly after a limited increase at the beginning of the instar. Feeding following a long period of starvation results in a marked increase in the volume of the gland which greatly exceeds that in normally fed specimens, not only in the first ovarian cycle but also in the later ones. The researcher indicated tthe possibility of resemblance with the adenohypophysis in vertebrates.

B. Scharrer (1956, 1958) examined the impact of ovariectomy on the ca in *L. maderae*, and observed an increase in the volume of the glands, the number of nuclei, and an increased nucleo-cytoplasmic ratio follow- ing the operation. Similar changes follow denervation of the ca (interruption of the nervi corporum cardiacarum). The researcher calls attention to the analogy with the neuroendocrine cycle in vertebrates. The succession of developmental changes in the ca in the adult bug *Oncopeltus fasciatus* was analyzed by Johansson (1958a) and that in the pupae of *Mimas tiliae* (Lepidoptera) by Highnam (1958b).

The disparity between ca from adult females which inhibited metamorphosis (induced 'supernumerary' larval moults) and those from adult males, which failed to produce this effect, was found by Fukuda (1963) in the silkworm *(Bombyx mori)*. This sexual difference in the activity of the ca persisted after gonadectomy, after both gonadectomy and reimplantation of the gonads of the opposite sex, and even in ca which were transplanted into another specimen after the beginning of the pupal period. The activity of the female ca was not affected by decerebration at the beginning of the pupal period.

Akahira *et al*. (1967) examined the effect of the ca in 32 species of stingless bees (genera *Cephalotrigona, Trigona, Scatura, Plebeia, Partamona, Nannotrigona* and *Melipona*) and two subspecies of the honey-bee *(Apis mellifera adamsonii* and*A. m. ligustica)* and observed their biometry (size, number of nuclei, volume, etc.), and histological characteristics. They agreed in many characteristics, but differed in others. In general, diversity was less in species of the same genus than in related genera.

Secretory activity of ca was observed histologically during sexual maturation and imaginal diapause in *Leptinotarsa decemlineata* and *Gryllus domesticus*. Changes in the size of the ca were observed in association with different stages of glandular secretion. In both species the secretory cycle was further classified into three periods – a pre- secretion stage, a stage of nuclear activity and a stage of production and release of the secretory substance. The ca of constrated females displayed hypertrophy and accumulation of hormone, which in later stages could result in degeneration.

A process of the volumetric measurement of the ca was expained and used to analyse the relation between the size of theca and their activity, determined by the *Locusta* colour test. A close correlation was found. Lea and Thomsen (1969), however, found no correlation between the size of the ca, their activity and egg maturation. Activity was judged by the neurosecretory cell nuclear test.

Ultrastructure

Electron microscopic examination of the ca and its functional changes began with the work of B. Scharrer (1961, 1962) on adult females of *Leucophaea maderae*. The secretory cells interlock by means of long cytoplasmic processes. Their cell membranes are folded in a characteristic way in inactive glands. In active glands the cytoplasm of the cells becomes more abundant, the number of mitochondria and ribosomes increases, an endoplasmic reticulum and very numerous granular inclusions appear. The gland is encased by a thick membrane from which processes of varying thickness branch off and penetrate the spaces between the secretory cells. These processes form a 'stroma', containing denser bodied, through which the neurosecretory axons of the nervus allatus penetrate the ca. The 'stroma' seems to be continuous with the adjacent cc. The researcher suggests that such process serves as a pathway through which food and other chemical are exchanged between the gland and the haemolymph, including diffusion of the hormone.

Subsequently, electron microscopic studies of various insect species of different orders have been published with reference to the secretory activity of the ca, on the original elaboration, storage and release of the 'C body', a specific cellular product in the ca of *Leucophaea maderae*. C. bodies are found or hexagonal structures of varying size, composed of highly electron dense material. They originate in the Golgi zone and are slowly transported to various intracellular and extracellular compartments affording them access to the haemolymph. From their incidence, distribution and fat in the insect organism, it was assumed that C bodies might have a bearing on the secretory function of the ca. Special attention was paid to the acellular stromal matrix, which seems to act as a vehicle facilitating the transport of C bodies to the haemolymph. C bodies might correspond to the 'osmiophilic bodies' described in the ca of the adult *Bombyx mori* female.

Number of research of papers on the ca of *Locusta migratoria*, their activity and the impact upon the glands of the A, B and C nsc of the pars intercerebralis was published by Joly and his co-workers. Abundant ergastroplasm was observed in actively secreting cells, which disappeared from cells in the non-active phase while the smooth endoplasmic reticulum showed conspicuous accumulation.

With reference to the previous discovery, Panov and Bassurmanova (1970) found an entirely different structure in the ca of the bug *Eurygaster integriceps*. Here, the smooth-surfaced endoplasmic reticulum was entirely absent in inactive cells, which were characterized by diminished size, reduced cytoplasm and numerous ribosomes. The most pronounced feature of active cells was a large nucleolus formed of hollow spheres like the one observed by Blazsek *et al*. (1973) in pgl cells of *Galleria mellonella*, while in inactive cells it was irregularly shaped with small vacuoles. The researcher discussed the possible functional significance of these changes in connection with similar observations in vertebrates (in the egg cells of *Protopterus ethiopicus* and in certain blood cells and liver cells of the rat). In addition, ultrastructural changes in the ca cells of specimens infected by larvae of the parasitic fly *Clitimyia helluo* were described (Panov *et al*., 1972).

The ultrastructure of the ca was closely scrutinized by Deleurance and Charpin (1972) in cavernicolous beetles of the family Bathysciinae, by Odhiambo (1966a, b) in *Schistocerca gregaria*, by Waku and Gilbert (1964) in *Hyalophora cecropia* and by Fukuda *et al*. (1966) in *Bombyx mori*. The cà of *Calliphora erythrocephala* female adults were examined ultrastructurally by E. and M. Thomsen (1969), with reference to hormone production. It was acccceoted that the hormone or its precursoe originates in the smooth endoplasmic reticulum (which was found to be very abundant at the stage of incipient activity) and that it dissolves in lipids, finally forming fat droplets which are thought to be gradually extruded through the surface of the ca. The fat droplets seem to correspond to the osmiophilic bodies of Fukuda and the C bodies of B. Scharrer. An exhaustive comparative electron microscopic study of the ring gland in the larva and of the cc-ca complex in the adult female of *Drosophila mellanogaster* was carried out by King *et al*. (1966).

Embryogenesis

As opposed to earlier belief that the ca was of mesodermal or even of endodermal, origin (cf. Nussbaum, 1889), their ectodermal origin now seems to be confirmed. Mellanby (1936) proved that they originate as ectodermal invaginations in the area between the mandibular and maxillary pleurites in *Rhodnius prolixus*. The original paired rudiments fuse into a single median body in Hemiptera towards the end of the embryonic period. Abnormalities due to incomplete fusion of the two embryonic invaginations, sometimes even showing two separate bodies, were often discovered by the reesarcher in the plant-feeding lygaeid bug *Oncepeltus fasciatus*. An increased number of such aberrations, amounting to 20 per cent, was observed in animals kept at higher temperatures. It is suggested that they are the result of a selective influence on the process of fusion exerted by the increase in temperature. Similar changes have been observed in *Pyrrhocoris apterus*. The same origin was found by Pflugfelder (1937a) in Carausius morosus where the glands retain their original vesicular shape both in nymphs and adults. The embryogenesis of the ca as a part of the ring gland was elucidated by Poulson (1945a, b) for *Drosophila*. Their origin from ectodermal invaginations in bugs is also stated by Wells (1954).

Phylogenesis

Ca has been discovered in all groups of metabolous insects (Ptery- gota) where they have so far been looked for. In Apterygota, they are fairly well developed in Japygidae and in a more primitive stage of development, in a lateral-position closely connected with the mandibulo-maxillar area of their origin, in *Cteno- lepisma*. A distinct juvenilizing effect of the glands of *Ctenopisma* implanted into *Cecropia* pupae has been observed. Keeping in view the earlier suggestions of the possible homology of the ca with the 'crops jugales' in Thysanura must be considered. It is probable that the original function of the ca was the activation of the ovarian follicles and that only gradually did they gain their function in larval development which led to the origin of Pterygote insects. The structure and development of the ca in the larvae of higher Diptera is quite complex. Here they form a part of the ring gland (1) the dorsal fusing of the originally paired ca; (2) the fusing of their nervi allati with the neighbouring pericardial glands; (3) the strengthening of the lateral arms and the dorsal part of the gland with tracheal trunks; (4) the bending of the ring with its upper part forward by a backwards shift of the brain. As pointed out by Cazai (1948) the formation of the retrocerebral endocrine system corresponds in a remarkable way to modern views on the phylogeny of metabolic insects based on the system of Martynov (1937) and Jeannel (1949).

The direct effects of JH

Ever since Wiggleworth (1935, 1936) discovered JH and demonstrated that it inhibited metamorphosis and activated the ovarian follicle cells, many rearcher papers have been written on various effects of the ca hormone and its different analogues on the most diverse structures and functions of the insect organism, on its various biochemical processes and on its chemical composition. Actually taking the entire insect organism as a whole, and not its functions, structures, chemical processes or components has been found to be independent of JH.

The inhibition of metamorphosis

As it has been already stated, implantation of the active ca at the beginning of the last larval instar partially or completely prevents metamorphosis and results in adultoid forms or supernumerary larval instars. These extra instars (giant larvae) may actually change into giant adults at the next moult. The removal of the ca by decapitation or extirpation in earlier larval instars is followed by a precocious metamorphosis resulting in miniature adults, preceded (in Holometabola) by miniature pupae. In most Hemimetabola only one instar occurs following allatectomy (e.g. Hemiptera), whilst in a number of the phylogenetically 'oldest' groups, such as Blattoptera and Phasmidae, two more moults appear following the operation, a special 'pre-adultoid' instar preceding miniature adults.

The suppression of metamorphosis connected with the positive effect upon larval (isometric) growth is the quite obvious and important effect of JH oneof the many feactures. Implantation of the c markedly accelerates the moulting process. For example, the intermoult period of the last larval instar in *Rhodnius* takes 24 days, whilst following the implantation of active ca at the beginning of the instar, the animal moults in 17 days. On this Wigglesworth (1936) based is original concept of the JH effect, applying Goldschmidt's theory of differential velocities (1923). He finally came to the decision that morpho-genesis could not be completed because of the acceleration to the moulting process. Such an effect really appears to exist but only till a limit in the case of the so-called progressive prothetely.

The effect of the JH on the growth and morphogenesis of internal organs

Sehnal (1965, 1968) in *Galleria mellonella*, fully accepted that JH affected the surface structures. The implantation of three complexes of brain-corpora cardiaca-corpora allata at the beginning of the last larval instar completely inhibited metamorphosis. Later implantation resulted in a series of transitions between the larval and imaginal shape of the brain and other organs.

Hlinak made the same discovery (1968) in the brains of a continuous series of transitions between a supernumerary larva and a normal adult in *Periplaneta americana*. Other similar results on the internal metamorphosis of the imaginal compound eyes were confirmed by Mouze and Schaller (1971) and Mouze (1970, 1971) who distinguished it from the metamorphosis of external structures, in which mitotic activity (and hence differentiation) depends on the moment at which an active MH concentration is reached. Only from that moment can JH action take effect, correlated with its prticular concentration in the haemolymph. On the other hand, in the MH-independent organs, the moment at which only a minimum active concentration is reached is decisive. The results are also confirmed by Riddiford (1972) in *Hyalophora cecropia*.

Dependding on corcumstances JH appears also to be capable of accelerating metamorphosis. At least this would be the most probable explanation of some of the observations of Wigglesworth (1948a), according to which the removal of the ca at the beginning of the last larval instar results in somewhat nymphoid adults. From this it would follow that the presence of JH in a concentration below the morphogenetical threshold is necessary for complete imaginal differentiation. On the other hand, the positive effect of the implantation of last instar ca on imaginal differentiation in earlier nymphs appears to have another reason.

The introduction of an active ca into an adult by implantation or parabiosis can result into a partial suppression of the imaginal characters, when an artificial imaginal moulting is induced simultaneously, either by parabiosis with the nymphs, or by implantation of active pgl, or by the injection of ecdysone or by exposing a part of the imaginal epidermis to the action of the larval hormone as in the epidermal bladder method.

The effects of JH on the indirect muscles of *Leptinotarsa* corroborated with concepts of all the other effects of the hormone. There is, however, an exception – the experimental results of the action of JH and JHa on the indirect flight musculature of the cricket *Gryllus domesticus*. Here the impact of the hormone is the exact opposite of observations elsewhere : the muscles ripen without JH and its administration causes their degeneration. The flight musculature persists in allatectomized females while the reintroduction of JH or JHa, at any time, causes its rapid degeneration. Castration of females in the last larval instar affects neither maturation nor degeneration of the muscles. In the last larval instar the action of the hormone is the same as in all other insects. This is the only one of the yet expained paradoxical effects of JH. At present, the probable theory would be that the presence of the hormone acts as a signal for nervous action causing flight muscle degeneration. Nervous inervention is evident since removal, or even mereimmoblization of the wings also causes the flight muscle to degenerate or inhibits their maturation.

Recent data by many authors, supplemented by the study of JHa effects, show that JH can unquestionably affect morphogenesis at any point from the initiation of egg segmentation up to adulthood. This is demonstrated most convincingly by the action of JHa on embryonic development (Slama and Williams, 1966b; Novak, 1969; Riddiford, 1969-197ia, Matolin, 1970; Rohdendorf and Sehnal, 1973) (see p. 219). Some papers describe th effects of JH on the pupal instar (Riddiford, 1972; Riddiford and Ajami, 1972) and its sensitivity to the administration of JH and JHa up to the next ecdysis. Various other authors, e.g. Sehnal and Meyer (1968), arrived at similar conclusions. Gametogenesis can likewise be inhibited hy JH at any of its stages (see p. 241).

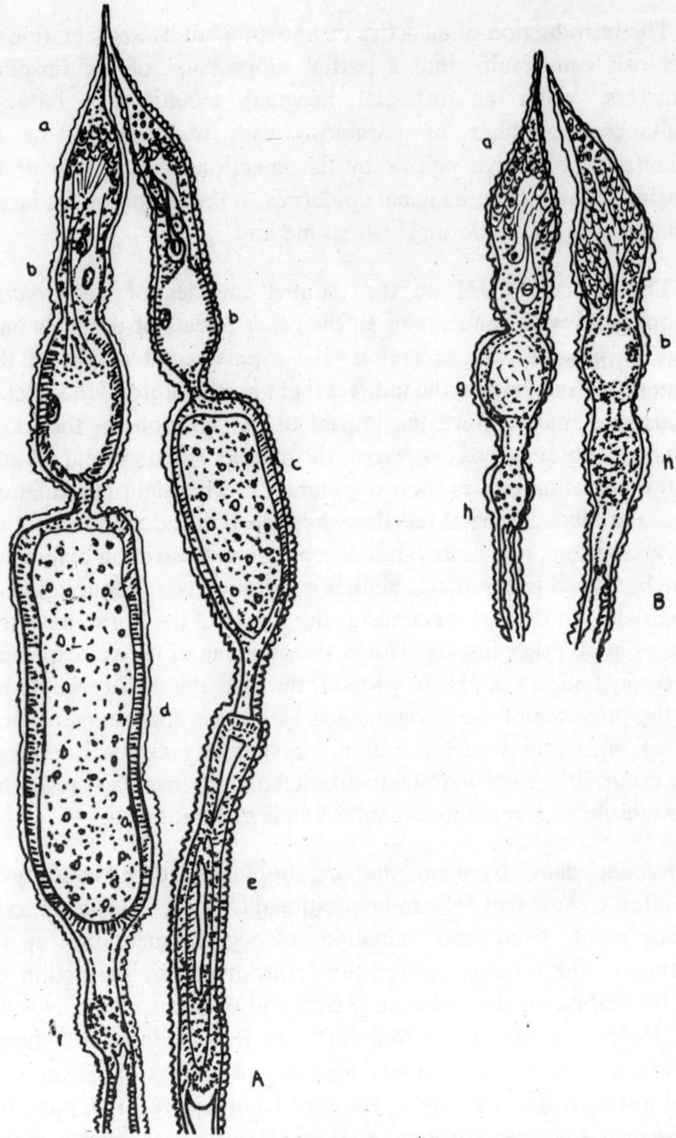

Fig. 7.15. Effect of JH on egg development in *Rhodnius prolixus*. A, Two ovarioles of normal female with active c. allatum; B, Two ovarioles of allatectomized female. *(a)* germarium with nurse cells, *(b)* oocytes nourished via nutritive cords from nurse cells, *(c)* oocyte nourished solely by follicular epithelium, *(d)* egg almost fully developed, *(e)* empty follicle, *(f)* last remains of corpus luteum, *(g)* dead oocyte, *(h)* residues of dead oocytes and follicles undergoing autolysis.

The influence on ovarian follicular cells

The effect of the ca hormone on the development of eggs in the ovaries was demonstrated in one of the initial researcher papers of Wigglesworth papers on JH. Wigglesworth (1935, 1936) ascertaned that the presence of an active ca is absolutely indispensable in *Rhodnius* for the ripening of eggs beyond a certain developmental stage, at which the ovarian follicle cells become active. Following allatectomy the growth of the eggs stops at a certain stage because of the degeneration of the follicle cells. This effect may be avoided the simultaneous implantation of an active ca from another specimen. Larval and imaginal ca show the same activity in this respect. Results similar to those with *Rhodnius* were obtained by Joly (1945a, b) with the beetle *Dytiscus*, by Weed-Pfeiffer (1945) with the grasshopper *Melanoplus differentialis*, by B. Scharrer (1946) with the cockroach *Leucophaea maderae*, by E. Thomsen (1952) with the blowfly *Calliphora erythrocephala* and by other reseachers with other groups of insects. The research on *Calliphora* was further developed by Possompes (1955), according to whom extirpation of the ca suppresses the ripening of the eggs only when performed in the adult. But it is still unknown whether or not there is at least a partial regeneration of the ca after its earlier removal. In the higher Diptera, the adult ca are quite different from those of the larvae (the dorsal part of the ring gland) so that their development may not be prevented even by the complete removal of the larval structures. The necessity of JH for ovarian development has been found in most species examined in this respect, although there are species where no JH control exist, such as *Carausius morosus*. *Bombyx mori*, etc. This independence of the development of the eggs seems to be a phylogenetically secondary adaptation connected with a prticular type of egg-laying : for example, in *Bombyx* all the eggs are deposited in a short time, not long after the imaginal moult. While in *Carausius*, they are laid onr at a time at long intervals.

The presence of a special gonadotropic ca hormone has been suggested by various authors. As emphasized by Pflugfelder (1952, 1958), there is no definite proof in favour of this conclusion. There are, on the other hand, significant arguments in support of one hormone theory. One of the most important of these is the fact, already mentioned, that the effects of the imaginal glands can be reproduced by the glands of any larval instar not only of the same species but also of any other insect order. Also, conversely, the imaginal ca of any insect species can replace the larval ca in any insect (Pterygote) species. For instance, the ca of adult female *P. americana* in the process of ootheca formation can prevent imaginal differentiation in the last larval instar of *Oncopeltus fasciatus*/. The histological evidence also lends no support to the theory of two or more hormones. The male gonads are in all cases independent of the presence of the JH. The fact that one and the same substance is responsible for either effect has been confirmed with juvenoids.

The influence on reproduction

The problem of the endocrine control of reproduction has been dealt with by many researcher and reviewed by Doane (1962), by Highnam (1963) and by Nayar (1964). Nervous control of egg maturation by specific neurones along the ventral nerve cord was demonstrated by Engelmann (1964) in pregnant females of the viviparous cockroach, *Leucophaea maderae*. These nuerones are supposed to inhibit the ca by influencing specific regions of the brain. The researcher assumes that a humoral factor released by the brood sac affects these neurones, but his experimental evidence does not seem to prove this. The role of hormonal factors in the control of ovarian development in *Schistocerca gregaria* was investigated in a series of papers by Highnam (1962), Highnam and Lusis (1962), Lusis (1963) and Highnam, Lusis and Hill (193a, b). In *Locusta migratoria* castration in the last larval instar does not seem to affect the cellular volume of the ca in either males or females. Allatectomy in a highly autogenous strain of *Aedes taeniorhynchus* inhibited egg maturation when carried out within one hour after adult moulting; yolk deposition was stimulated by the implantation of one pair of active ca. Similar results were obtained in other aedine mosquitoes.

Some researchers have recently reported quite a different type of JH and JHa action. JH administration in the early phases of gametogenesis not only does not promote the process, but, on the contrary, inhibits it. The sensitive period for this inhibitory effect is, in general, earlier than for follicle cell stimulation. In some of the insects studied it comes at the beginning of the adult stage, in others in the pupal instar, or (in Exopterygota) in the last larval instar. It includes the whole period of gametogenesis, from the first oogonia and spermatogonia up to the mature sex cells. In principle, this is the same

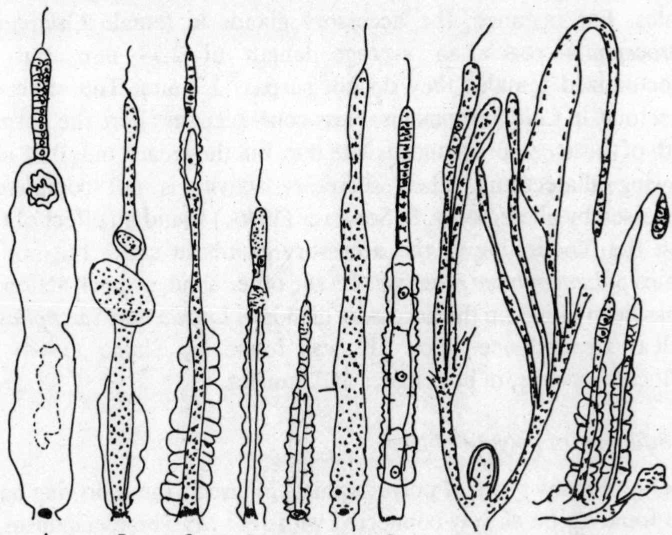

Fig. 7.16. Ovarioles of *Thermobia domestica* after juvenoid administration to preceding instar 20 to 50-days previously.

'inhibition' of morphogenesis as observed in any other phase of ontogenesis. As with embryogenesis, the results of such inhibition of growth can differ considerably, depending on to the time of administration of the substance and the duration of its effect. Various kinds of deformation of the ovarioles mostly result.

Many researecher papers have been published on the ca and JHa endocrine control of oogenesis and of reproduction in general. Whatever the conclusions reached by individual reseacher, the observed facts are all in agreement with the original explanation by Wigglesworth (1936) of the effect of JH on the ovarian follicles in *Rhodnius prolixus*.

The influence on the accessory sex glands

The necessity of JH for the functioning of the accessory glands in male *Rhodnius prolixus* was illustrated by Wigglesworth (1936a) in his first researcher paper on the ca hormone. Since that time a similar dependence of the accessory glands on JH has been shown in many other insect species, not only in males but in some cases also in females. For instance, the accessory glands in female *Calliphora erythrocephala* reach an average length of 2.58 mm, but in allatectomized females they do not surpass 1.7 mm. The effect of allatectomy in *Calliphora* males is less cons- picuous : here the normal length of the accessory glands is 1.18 mm, but they reach only 0.91 mm following allatectomy. Their secretory activity is not completely suppressed by allatectomy. B. Scharrer (1946c) found an effect of the ca on the functioning of the accessory glands in males but not in females of *Leucophaea maderae*. On the other hand, no such action of JH has been found on the accessory glands of *Lucilia* and *Sarcophaga*. A clear dependence upon JH was found by Slama (1964) in *Pyrrhocoris apterus*, in both males and females.

The influence of polymorphism

Out of many types of polymorphism in insects the following have been found to be closely connected with JH : *(a)* The mechanism of *caste polymorphism in social insects* has been found to bear some relationship to JH activity, as for example in ants and in termites. The latter authorsoriginally presumed the existence of a special, third ca hormone responsible for soldier production if administered to larvae or pseudergates at the 'right time'. This possibility was ruled out by Luscher (1969), however, who showed that Roller's synthetic JH also induced the same effect. Soldier differentiation requires much larger JH doses than those needed for the normal, i.e. juvenilizing effect. *(b) Phase polymorphism in locusts*. The infleunce of JH on the development of *Locusta* and *Schistocerca*, has been studied by Staal (1959, 1961), L. Joly (1960), Kennedy (1956, 1961), Loher (1960) and others. Their findings point to a complicated dependence of phase on JH activity. As confirmed by Staal, the implantation of active ca into young larval instars induces the green colouration typical of the solitary phase even in environmental conditions which would otherwise produce the gregarious phase. The adaptive value of either phase in an environment which induces this particular type of development has been emphasized by Kennedy (1961). The existence of a volatile factor acting through antennal chemoreceptors seems probably on the basis of the results of Loher (1960) who was the first entomologist to discover a susbtance accelerating sexual maturation in adult male locusts.

More information on this subject have been obtained by Joly and his co-workers. *(c) Seasonal polymorphism in aphids*. A very complicated type of polymorphism is encountered in aphids. There is a strict seasonal alternation of forms with a pre-ponderance of parthenogenetic generations. The polymorphis forms differ with regard to the occurrence of wings, the presence or absence of vivipary and in a number of other, less marked, characters, both morphological and physialogica. Even though we are unfamiliar with the the various mechanisms involved, there are definite suggestions that the metamorphosis hormones and perhaps other chemical factors play a part in the determinatio of the character of the various forms.The apterous virginoparae of *Megoura*, for example, appear to be produced by the neotenic effect of increased JH activity. This activity, however, seems to take place during the embryonic period through the action of the maternal endocrine system on the ca of the older embryo. The still earlier activation of the embryonic ca by the maternal body appears to be the cause of differentiation of the oviparae. Similar conclusions were reached by Johnson and Birks (1960), with an Australian aphid species *(Aphis craccivora* Koch.), who assumed that all aphids started their development as presumptive winged forms which may be diverted from this path of development by various stimuli in the determination period between the late embryonic stage and the second larval instar. These researchers assumed that the results they obtained, could be applied to to other types of insect polymorphism.

The effect of JH on aphid morphogenesis was examined by Holman (1973). In a detailed analysis he found that the statistical regression expression the relationship between the length of individual parts of the body and total body length or the breadth of the head capsule differed relatively little in the Ist to IVth instar, but displayed a significant shift between the IVth instar and the adult. The regression curve corresponded to the total changes which occurs in any part of the body during metamorphosis. In most cases the absolute value of the shift was also dependent on body size. In supernumerary instars and juvenilized adults, this imaginal shift was reduced to various extents. The degree of JH effects can be determined from a comparison of these data with the average normal individual. Mathematical formulae for the statistical evaluation of these effects are given. *(d) Alary polymorphism in bugs*. Alary polymorphism or the occurrence of divers degrees of wing development in one and the same species is a common phenomenon in Heteroptera. A theory has been elaboratied by Southwood (1961) based in Wigglesworth's findings (1960) on the hormonal control of this type of polymorphism. Brachyptery is assumed to originate on the principle of metathetely in some species, particularly in colder, mountain conditions. It may be classified as a juvenile character in adults following the normal number of nstars (five in Heteroptera). In these insects (such as *Gerris, Nabis, Bryocoris* and various *Tingidae*) macropterous forms are found mostly in warm localities. On the other hand, brachy- ptery is assumed to originate in other species on the principle of prothetely (paedogenesis). In species such as *Dolichonabis limbatus* and *Microvelia*, long-winged specimens appear in cold, mountain conditions whilst in the normal warm environment only brachypterous forms occur. This theory is yet unconfirmed as stated by its author. *(e) Sexual alary dimorphism in Lampyridae*. Davydova (1968) described the effects of ca implantation and extirpation on wing development and metamorphosis in both males and females of *Lampyris noctiluca* (Coleoptera). Allatectomy often resulted in considerable prolongation of life without metamorphosis. Prothetelic males (regressive prothetely) with smaller wings were sometimes obtained. The implantation of several ca induced one or two supernumerary larval moults in the females, without inducing any wing development. The absence of wings in the females, together with other neotenic characters, is obviously of earlier phylogenetic origin and is incapable of phenotypic modification.

The effect on regeneration

A remarkable decrease in regeneration ability was noticed by Pfulgfelder (1939b) in *Carausius morosus* following extirpation of the ca. In some cases, particularly when an amputation was effected after allatectomy, there was a complete loss of regeneration ability. Generally the degeneration of whole groups of cells was observed in the regenerating epidermis whilst knotty proliferations appeared in other parts. But no difference was observed in *Blattella germanica*, either in the extent of regeneration or in its effect on the duration of the instar concerned, irrespective of whether the extir- pation was carried out during the period of active JH concentration (Ist to Vth instar) or in its absence during metamorphosis.

Therelation between regeneration and the changes in volume of the ca were surutinizedin detail by O'Farrel *et al.* (1961) in *Blattella germanica*. The duration of the moulting process was only affected when muscle tissue was involved and the greater the amount of tissue affected, the longer the intermoult period. But the moment the regeneration was restricted to epidermal structures, no changes could be seen in the moulting cycle. No evidence is available of any nervous influence on the regulation of the moulting process and all the effects observed appear to be the results of self-regulatory processes. A histological examination of the volume of the ca and that of the so-called headlobes carried out at twelve-hourly intervals showed a close connection between the volume changes of the two glands. Regeneration has a positive effect on the functioning of both pairs of glands, which is particularly interesting from the point of view of the experiments of Bodenstein (1953) and Engelmann (1959) on the reciprocal effects of the ca and the pgl and their effect on the moulting process. Immediately after the moult following treatment which causes the regeneration, a hypertrophy of the ventral lobes and ca occurred. This hypertrophy, and the subsequent extensive regeneration seems to result in extra moults in the adult stage. The delay caused by the regeneration observed in the moulting process in the Ist larval instar of *Blatella germanica* is followed by a shortening of the subsequent instars so that the IVth instar moult occurs at the same time as that of the controls. This can also be assumed to be as a of the positive effect of regeneration on the functioning of the gland concerned.

The effect of metamorphosis hormones on the regeneration process in insects was explained in great detail by Boden- stein (1955, 1959) on the basis of his numerous experiments. He emphasized the need of MH for regeneration. It is only the degeneration of the prothoracic glands associated with metamor- phosis which makes the adult metabolous insects unable to moult. All the tissues of the imago retain their full capacity for regeneration, which occurs in the extra imaginal moulting process brought about by the implantation of active prothoracic glands or by parabiosis of the adult with a nymph. It is yet unclear as to what extent the observed effect of MH on regeneration is direct or, perhaps more probably, whether it is only indirectly brought about by inducing the moulting process. Accorrding to bodenstein the results obtained by him contradict those of Pflugfelder with *Carausius morosus*, and attempts to find a common.

Penzline (1965a, b) has comprhersively analyzed of regeneration and the ways in which it can be influenced hormonally. He illustrated that all three metamorphosis hormones were needed for normal regeneration and that it was inhibited by allatectomy. Another detailed survey was published by Needham (1965), who suggested that JH was necessary for regeneration processes to a limited extent in adult insects, e.g. the healing of integumental wounds and the regeneration of nerve fibres and internal organs. Thus capacity can be greatly enhanced by the transplantation of wing discs and leg fragments, etc., and by the implantation of active pgl. Rinterknecht (1964) found that allatectomy did not inhibit the reconstruction of the integument in *Locusta migratoria*, but caused a roughly two-fold dealy and that it similarly slowed down the mitotic activity of the regenerating tissue. Normally allatectomy is supposed to speed up ageing.

The influence on tumour growth and other alterations in histogenesis

Including to the morphological effects earlier mentioned, JH has atremendous impact on histogenesis. The first detailed analysis of the histological effects of ca extirpation was carried out by Pfulgfelder with *Carausius morosus*. He noticed numerous changes of a pathological nature following alltectomy carried out in earlier instars. He distinguished tissue degenerations and tissue proliferations both of which often occurred in one and the same tissue. The most reoccuring pathological changes are : *(a)* Degeneration sometimes connected with a partial histolysis of fully developed tissue and often connected with a subsequent partial or complete phagocytosis. Simultaneously, proliferations usually appear as knotty structures scattered throughout the tissue, which sometimes contain single differentiated cells in the form of giant muscle cells. *(b)* Degeneration of the nervous system and the adjacent mesoderm and glial cells in which the formation of large irregular cysts often occurs. *(c)* Degeneration of the fat body during which vacuoles appear in fat cells, which resemble fat droplets but do not stain with osmium tetroxide and therefore probably do not contain either fat or any lipoid substance. It would be interesting to determine to what extentdefine the limit of JH deficiency syndromes agree with the alterations in the fat body of adults of other species, which are not connectred with patholotgical changes in other tissues. The prolonged effect of allatectomy is encapsulation of the pathologically changed cells by lymphocytes which often leads to the development of large cysts. Inside these cysts large concretions often appear which show concentrical layers. *(d)* Degeneration features are also, often observed in the Malpighian tubules in which cysts and concretions, similar to those in the fat body, appear. *(e)* Accumulations and proliferations of haemocytes (lymphocytes) often occur. These may give rise tonumerous knots of undifferentiated cells, which resemble those of spindle cell sarcoma in Man. *(f)* Proliferations of a carcinomatous nature are observed in the epidermis, gut and oviducts.

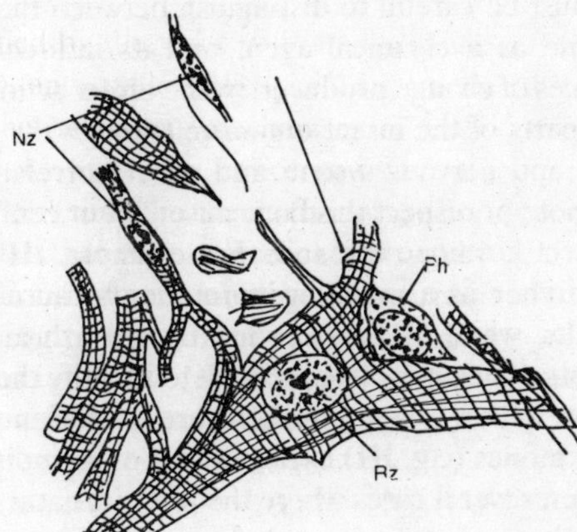

Fig. 7.17. Malignant growth of muscles in *Carausius morosus* following allatectomy. *Rz* – giant muscles cell; *Nz* – normal muscle cell; *Ph* – phagocyte (haemocyte).

Many changes occur following the implantation of extra ca (from adults into larvae), in, for example, the ovarioles, oviducts and other organs and tissues. But they are, certainly different from the changes produced by allatectomy.

Various modifications in tissue differentiation frequently occur during regeneration, particularly when a piece of embryonic tissue is implanted whilst there is disturbance in the hormonal balance, as shown by Pflugfelder (1948). He attempted verify this by the fact that young cells are more easily affected by external stimuli than older ones. Tumorous malignant growths have often been observed following the cutting of various nerves, such as the nervi corporum cardiacarum.

Schneiderman and Gateff (1967) set up clones procured from a new mutant in *Drosophila melanogaster*, which they called *lethal malignant brain tumour*. The mutant imaginal discs grew considerably but did not respond to react the metamorphosis hormones. The mutant cells grew rapidly and invasively, killing the hosts, and behaved like true malignant tumours. They were cultivated for 35 transfer generations and further clones were secluded from single cells.

The indirect effects of JH

One must be careful to distinguish between the direct effects of the hormone as a chemical agent and its indirect effects, i.e. the consequences of change produced by the direct action of the hormone in various parts of the insect organism through the mediation of the circulatory and nervous system and other correlation mechanisms. Failure to note or respect this fact caused lot of confusion in the early days of insect hormone research. For example, JH was regarded by some researcher as a moulting hormone, because it induced extra larval moults, while MH was called the growth and differentiation hormone, etc. Since it is often difficult to identify the actual character of hormone activity, many authors are at variance as to whether specific hormones (e.g. JH) exert a direct or an indirect effect. There are, however, several cases where the indirect nature of the JH effect is confirmed.

The effect on moultin

The researchers, who were the first one to deal with the ca hormone assumed it to be a moulting hormone. Depending on the fact that, after the implantation of active ca into a last instar larva, one or several extra moults and supernumerary larval instars occurred. Further investigations have, however, shown without doubt that the moulting process is directly influenced by the prothoracic gland hormone (MH), and the remarkable effect of JH is actually the indirect effect of prolonging the action of the pgl which are normally histolysed during or immediately after metamorphosis. The seemingly contradictory observation of Bodenstein (1953a-e) that allatectomy towards the end of the last-larval instar in the cockroach *Periplaneta americana* may result in preservaton of the pgl and thus cause supernumerary moults can also be explained on this basis. As demonstrated by O'Farrell *et al*. (1953, 1956), the amputation of legs or a more extensive injury to other parts of the body may have the same effect as ca extirpation. This, as noted above is an effect of the regeneration process, perhaps improved by selection into a self-regulatory mechanism, in the case of insects with a well-developed autotomy, and has nothing to do with the action or absence of ca. The indirect nature impact of JH on moulting is quite obvious from the fact, amongst other things, that neither the normal nor the experimentally induced action of the ca in adults causes any sign of moulting, whilst this may easily be induced in the adult, by the implantation of an active pgl.

The effect on the silk glands of caterpillars

It is generally accepted that in many Lepidopterous larvae the labial glands are almost inactive during the first to penultimate larval instar whereas they produce a large quantity of a silky material for cocoon spinning in the last larval instar. The situation is just the reverse in sawflies such as *Cephalcia* (Pamphilidae) where the silk glands are active throughout the larval instars producing the 'excrement webs' in which the larvae live gregraiously, and pupation takes place in cells in the soil without a cocoon. In each contact the action of the glands is connected with the presence of JH. The hormone, however, does not affects the silk glands directly, but affect their morphogenesis in a very general way. Whether its presence affects the function of the glands positively or negatively depends on genetic, species-specific factors.

The effect on instincts

The effect of JH on cocoon spinning in *Galleria mellonella* has been explained as a special case of affecting instinct. The larva of this species spins a cocoon before each moult. This cocoon is, however, different before the larval moults from that before the pupal. The larval cocoon is in the form of a tube open at each end whilst the pupal one is spindle-shaped and completely closed. It was shown by Piepho (1950b) that the implantation of ca at certain times before moulting caused the larva to produce cocoons intermediate between the larval and the pupal types, depending on the time of implantation. This effect cannot be expressed as a direct one on the instinct of cocoon spinning but, undoubtedly, as an indirect one through the morphogenesis of the brain. By preventing the morphogenesis of the imaginal brain at the required time JH determines whether, and to what extent, the material basis for a given instinct is developed, i.e., the corresponding nervous (cerebral ganglion) structure. A detailed examinationof this and similar effects will no doubt make possible an exact localization of various imaginal instincts in insects, especially when these differ from those in the larvae, as for instance in the case of insects with aquatic larvae.

The question of hormonal influences on sexual behaviour in various species of the order Saltatoria has been described by Loher (1964). In some species, allatectomy influences the males, while in others the females are affected, and there are species which are not influenced by the ca at all *(Gryllus campestris)*. No effect on sexual behaviour of either allatectomy or the implantation of additional ca was found by Roller, Piepho and Holz (1963) in *Galleria mellonella*. On the other hand, Beetsma, de Ruiter and de Wilde (1962) report that the injection of *cecropia* extract shifts the balance between the photopositive and photonegative orientation tendencies to larvae of *Smerinthus ocellata* towards a positive response. The implantation of active prothoracic glands resulted in the reverse effect.

Some JH effects are still debatable. However, only affects immediately connected with the morphogenetic processes of stimulation of larval growth, inhibited derepression of more advanced characters and suppressed degeneration can be regarded as direct. With the discovery of juvenoids (JHa), we have another type of effects which may not be the direct action of the hormone, but may be resulting from the side effects of molecular structure or an effect connected with experimental hormone supply (e.g. an effect on membrane conductivity, heart rhythm and also, perhaps, interruption of the diapause). More information on this group of effects is required before they can be properly difined.

No difference has been discovered till now between the effects of the actual ca hormone and the various JHa, apart from their specificity form some groups of insects. This is not related to the mechanisms of action of the given JHa, but to their capacity for penetrating the integument, their resistance to various haemolymph enzymes, etc. All JHa effects can thus also be ascribed to JH and *vice versa*. This is specially significant in cases in which ca application is very difficult or even impossible, e.g. during early embryogenesis.

It is often difficult to differntiate between the morphogenetical and the physiological effects of a hormone as well as between the direct and indirect effects. All morphogenetical effects are necessarily based on physiological (biochemical) ones,while most of the physiological effects can, sometimes, cause morpho- genetical changes.

In the earlier edition of this book (Novak, 1966) there were doublt about certain results of JH action among its indirect effects. Since then, however, we have acquired much information about the mode of action of this principleand my original assumpptionon the primarily morphogenetic action of JH has been confirmed by dozens of new findings. Therefore, there is to be full justification for transferring most of the physiological and biochemical effects from the direct to the indirect group including, for example, all metabolic effects, together with oxygen consumption, the effect on fat metabolism and the fat body, etc.

The effect on the total metabolism (oxygen consumption)

It was indicated by Pflugfelder in as 1941 (on the basis of his work with ca implantations and extirpations in *Carausius morosus*) that JH is a factor which generally favours metabolism. This conclusion was later confirmed by the experiments of Weed-Pfeiffer (1945a). From her experiments on the effects of allatectomy in adult *Melanoplus differentialis* females, She discovered that the ca can exert a positive effect on basal metabolism. She observed a connection between the presence of the ca and the consumption of food reserves in the fat body as well as an effect of JH on the increase in weight of the fat-free dry-matter and water. Further confirmation was obtained by E. Thomsen (1949), who found that allatectomized female adults of *Calliphora erythrocephala* show a 19 per cent decrease in oxygen consumption. A similar dependence on the presence of JH was discovered in males. Experiments carried out under the same conditions using castrated females. (Thomsen, E, and Hamburger,

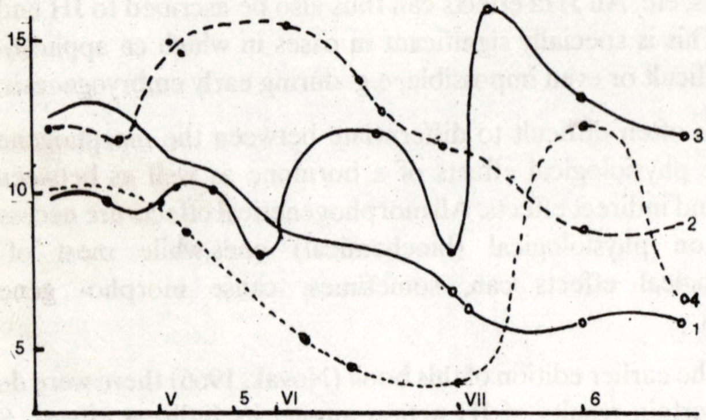

Fig. 7.18. Blood sugar concentration in normal and alltectomized nymphs of *Carausius morosus*. 1, 2 – concentration of sugar in the blood; 3, 4 – concentration in the tissue; 1 – controls; 2 – allatectomized; 3 – controls; 4 – allactectomized. Abscissa – time (Roman figures – number of the instar of the controls, Arabic figures – number of the instar of the allatectomized specimens). Ordinate – concentration percentage of sugar.

1955) apparently suggest that this may be a direct, and not an indirect effect on metabolism and not, as earlier supposed by Pflugfelder (1952), caused by oxygen consumption of the developing ovaries. Pflugfelder himself (1952) observed an increase in oxygen consumption in *Carausius morosus* following the implantation of ca and a decrease following their extirpation. Both effects were, however, rather weak and only temporary. L'Helias (1956a) noticed a decrease in oxygen consumption in the same species following allatectomy. An even greater decrease was observed when the cc were removed at the same time. L'Helias attempted to explain this as being connected with the accumulation of glycids and mineral phosphorus in the tissues and with the loss of energy caused by the breaking down of esters due to the increase in alkaline phosphatases.

A progression of experiments on the oxygen consumption of normal and castrated adult females of *Pyrrhocoris apterus*, implanted with ca appears to indicate an indirect action of these glands on metabolism. It was noticed that castrated females do not show any increase in oxygen consumption compared with normal controls following ca implantation, very probably because of the increased share in the total metabolism of the ripening ovarian follicle cells which do not grow in the absence of the ca. Examination of the effect of ca implantation into the last larval instar of the same species gave corresponding results. A direct connection was discovered between the extent of the induced morphological changes and therate of oxygen consumption. It was concluded from both series of experiments that the JH-effect on oxygen consumption is an indirect one, depending on the increase in the amount of metabolically active tissue. The direct effect of JH is probably concerned with protein synthesis.

But, Sagesser (1960) claims to have theory an increase in oxygen consumption after ca implantation in the cockroach *Leucophaea maderae* even in castrated females and concludes, in agreement with E. Thomsen, that JH affects metabolism directly. His point of view based on the theory of two ca hormone appears tibe incorrect.

Results which seem to contradict to some extent the above theory, are reported by Samuels (1956). He noticed a considerable increase in the oxygen consumption of the isolated thoracic musculature of *Periplaneta americana* adults of both sexes 2 to 3 months after allatectomy. Samuels tried .to explain the differences observed in oxygen consumption in the whole animal and in the isolated tissues by assuming that the glycide reserves, accumulated in the tissues decause of allatectomy, and thus prevented from being used inside the organism, are freed and used in the isolated muscle, thus increasing its oxygen consumption.

Slama (1960) examined the correlation between oxygen consumption and postembryonic evolution in various groups of insects; he found a typical increase at the beginning of each instar and a decrease in the second half of each larval instar. The period of metamorphosis was characterized by a typical U-shaped curve in Exopterygota also. A similar correction between oxygen consumption and ovarian development was found in adult *Pyrrhocoris* females, in which typical O_2 consumption cycles werre observed in connectionwith the ovarian cycles. The cycles were controlled in a typical manner by AH and JH. The correlation was less obvious in males, in which metabolism is low and displays no clear cyclicity (Slama, 1964b). Experiments with isolated body fragments showed that JH regulated oxygen consumption indirectly, by stimulating the growth, function and metabolism of specific parts of the body (Slama, 1964c). A similar close correlation was shown to exist between oxygen consumption and the degree of morphogenetic changes in *Galleria mellonella*. Sehnal (1966) performed a systematic study of the effect of JH on oxygen metabolism during larval and pupal development in the same species. No effect of ca implantation was observed in the penultimate larval instar, but implanttation in the last larval instar was followed by a dfinitet increase in metabolism in specimens which transformed to supernumerary larvae.

Effects on protein and nitrogen metabolism

The first research work on the ca effect on nitrogen metabolism was published by L'Helias (1956a), who observed the accumulation of amino acids in the tissues of *Dixippus morosus* following allatectomy. Janda and Slama (1964) studied the effects of JH on the body nitrogen, glycogen, fat and uric acid content in *Pyrrhocoris*. They were all proved to be indirect and the outcome of changes produced by the hormone on the growth process in particular parts of the body. Allatectomy, inhibiting ovarian development (and hence the utilization of reserve materials) resulted in enormous accumulation of these substances in the body. If cardiacectomy was performed at the same moment the production of reserve materials likewise stopped, so that compared with the controls no important differences were obsrved. In another study, Slama (1964) observed excessive accumulation of the haemolymph proteins following allatectomy or castration, which was diminished by ca reimplantation. Minks (1965) found differences between various protein fractions and free amino acid concentrations in the haemolymph of normal and allatectomized specimens of *Locusta migratoria*.

The two entomologists Bentz (1969), and Bentz Girardie (1969) observed changes in ovarian protein composition during egg maturation after both allatectomy and nsc electrocoagulation. Engelmann (1969) deterd the *de novo* production of a specific protein in the serum of adult *Leucophaea maderae* females treated with JH. The protein was absent from the in nymphs of either sex or in adult males, and it disappeared in allatectomized specimens. Engelmann, Hill and Wilkens (1971), using various JHa, found dependence of a similar female specific protein in *Schistocerca gregaria* and *Sacrophaga bullata*. Engelmann (1971) further analysed these effects in the original species *(Leucophaea)*, using different JHa, and obtained largely the same results. Since then, female specific proteins have been found in representatives of several insect orders. Engelmann (1972) later found that synthesis of this protien was blocked by actinomycin D. In agreement with this finding, JH also increased the synthesis of total sodium dodecyl sulphate phenol-extractable RNA, which stimulated specific protein synthesis in a completed cell-free system from egg- maturing females, whereas RNA from males or allatectomized females had no such effect. As stated by the researcher the information seems to support Karlson's hypothesis of hormonal control of the transcription of specific messenger RNA.

Effects on protein metabolism

The influence of JH and the ovaries on the amount of pupal and adult body fat was examined in *Musca domestica* by Adams and Nelson (1969), who observed a decrease in the rate of disappearance of the pupal fat body following both ovariectomy and allatectomy. The size of the adult fat body was not affected by allatectomy, but in ovariectomized specimens it was larger than normal. Baumann (1969) investigated the effect of several JHa on artificial bimolecular lipid membranes; after adding JHa he detectedgreater stability of the membranes, which varied with the kind of juvenoids and the type of membrane.

The influence on the fat body

It has been ascertained by Weed-Pfeiffer (1945a) that alltectomy causes an increase in the fat content which is only slightly affected by castration. Similar results were obtained by E. Thomsen (1942) with *Calliphora erythrocephala* and by Vogt with *Drosophila*. The latter researcher also observed reduction in size of the nuclei and nucleoli of the fat body cells which she explained as being due to the cytoplasm of these cells becoming filled with fat droplets. Day (1943a) also mentioned conspicuous histological changes in the fat body following extirpation of the ring gland in the flies *Lucilia* and *Sarcophaga*. It is not, however, very obvious from his experiments to what extent these changes were becuse of other components of the ring gland.

Gilbert (1966) examined the variation in the lipid content of the ovaries and fat body of *Leucophaea maderae* and found that lipid synthesis by a maturing ovary was enhanced by JH application *in vitro*. He gathered that JH might depress lipid synthesis in the fat body, with resultant mobilization of the reserve substances on behalf of the developing oocytes. There are no proof against the opposite possibility, however, which would fit everything else we know about the action of JH, i.e. that the action of the hormone consists in activation of the ovarian follicles. The metabolism of the ovarian follicles is much greater than that of the fat body, so they may attract reserve substances not only from the haemolymph but from the fat body also. Stephen and Gilbert (1970) also noticed that fatty acid composition in *Hyalophora cecropia* changed during metamorphosis and that it was affected by JH. Here again, the effects of JH are clearly indirect and caused by the different requirements of the differentiating tissue. Morohoshi and Kiguchi (1969), Morohoshi and Fugo (1972) and Morohoshi *et al*. (1972) studied various aspects of the effect of JH on lipid and other forms of metabolism in *Bombyx* (1969) and attempted to explain their findings on the lines of Morohoshi's hypothesis.. Again, however, there is no information which could not be discribed on the basis of the much simpler hypothesis formulated by Wigglesworth (1936) nearly 40 years ago.

The influence on the oenocytes

Reduction in the volume of the oenocytes and in the size of their nuclei with associated pycnosis was observed by Day (1943a) in *Lucilia* and *Sarcophaga* and by Vogt (1947) in *Drosophila*, following ring gland extirpation. It is difficult to decide the of changes resulting from JH deficiency and the absence of the other two metamorphosis hormones (AH and MH), which also originate in the ring gland. The possibility that the effect observed may be merely the result of inhibition of the moulting process must also be considered

The effect on the water balance

Including the neurosecretory cells of the pars intercerebralis and the cc, the ca are also said to produce an effect on water balance.. Allatectomy results in a reduction of the water content of the body, including that of the haemolymph. The effect of JH is particularly evident in castrated females where retention of the ca results in a considerable increase in the quantity of haemolymph with an associated conspicuous distension of the abdomen. In such individuals the amount of dry matter also increases. However, it is not clear exactly what effect is due to JH itself and what is because of the neurosecretion stored in the ca.

The effect on colour change

Quite obvious changes have been detected in the colouration of both larval and adult insects following experimental interference with JH production. Some of these are obviously of morphogenetical origin, such as the green colouration in the solitary phase of the migratory locust. Others are, undoubtedly because of to the brain neuro- hormone. There are, however, other changes which can be assumed to be due to JH, such as those causing the colouration of allatectomized stick-insects *(Carausius morosus)*. The colour changes are assumed to be caused by an accumulation of uric acid in the tissue together with a decrease in the purine, pterine and melanin contents. Changes in colouration were also observed in the same species following implantation of active ca. A marked morphogenetical effect of the ca hormone on the black pattern in the lygaeid bug *Oncopeltus fasciatus* has been written about.

Conforming the earlier findings on the colouration of *Gryllus bimaculatus* (1960), Roussel (1967) obtained several interesting results on the action of ca and the control of pigmentation in this species. Roussel reported very high individual variability in the response of females to the action of JH and found agreement between the control of melanin pigmentation in this species and the gregarious *Locusta migratoria*. The absence of the ca leads to an increase in black pigmentation, while only a small fragment of ca induced orange pigmentation. The researcher confirmed that the ca had a direct effect on the integument.

The cuticle colouration of the tobacco hornworm, *Manduca sexta*, was sanalyzed by Truman *et al*. (1973) as the basis for an ultrasensitive JH test. The presence of JH inhibited melanization of the cuticle in the larvae of this species. If the hormone was administered up to a given moment (between the 17th and 25th hour of the last larval instar), pigment formation depended on the absence of JH. This test has the same sensitivity as the wax test in *Galleria*, but can be experimented upon only two days after administration.

The effect on the mitochondria

An increase in the oxygen intake is stated by Clarke and Baldwin (1960) in mitochondriae preparations from *Locusta migra- toria* L. on a sodium succinate substrate, after the addition of ca. The increase was, however, merely detected in mitochondria prepared from *Locusta migratoria* adults, those from Vth instar nymphs. of *Schistocerca gregaria* showing a decrease in oxygen consumption. The addition of 2 : 4 dinitrophenol in physiological amounts, on the other hand, resulted in an opposite effect : it decreased the oxygen consumption of the mitochondria of adult *Locusta* and increased that of mitochondria of Vth instar *Schistocerca*. The addition of both dinitrophenol and ca at the same time increased the oxygen consumption of the preparations from both species. The researcher discuss the possibility of the ca controlling metabolism by their action on the mitochondria and the possible relationship between the effect of JH and those of the dinitrophenol.

But it is possible that these are indirect effects, and that they depend on the type of mitochondria on which JH acts. JH has no effect on imaginal (JH-independent) cells, but may act on the structures of larval (i.e. JH-dependent) cells. This also seems to be indicated by the lack of differences between the alary muscle mitochondria of normal and allatectomized *Locusta migratoria* (Joly *et al*., 1969), findings by Novak and Slama (1962).

The effect on pheromone production

It has been confirmed by Loher (1961, 1962) that JH controls sexual maturation and the corresponding changes in colouration and mating behaviour in the locust *Schistocerca gregaria* males in a very interesting way. The maturation of the adult males of the gregarious phase in this species is marked by three distinctive features : colour change from the light-brownish grey and white of the freshly moulted adult to the yellow of the mature specimen; a specific sexual behaviour pattern; and the production by the epidermal cells of pheromone which acclerates the maturation process in young adult males in the immediate vicinity of the mature specimen by an olfactory effect on their chemoreceptors. None of these indications is demonstrated allatectomized males. They are however, induced by the implantation of active ca from a mature specimen.

Th same type of pheromone was discovered by Barth (1962) in the cockroach *Byrsotria fumigata*. The virgin female of this species produces a volatile sex attractant which stimulates the courtship behaviour of the male. The production of this pheromone is also controlled by JH. It is not manufactured by allatectomized specimens, but its production can be induced by the implantation of active ca from another specimen. In this species no other role of JH in the reproductive behaviour of either of the sex was found

More particular JH (or ca implantation) effects in various indectswere observed by researchers. All of them were definitely indirect, or were perhaps, in some cases, caused by extrinsic neurosecretion contained in the ca. For example Baumann (1968) found that JH clearly increased cell membrane conductivity in the salivary glands of *Galleria mellonella*, causing strong depolarization of the membrane. The source of the hormone was *cecropia* oil. The same oil extracted from allatectomized males was inactive. Here again, the action of JH can be taken to affect prolongation of the activity of similar larval (JH-dependent) cells about to degenerate, or which have already actually started to degenerate, and that this also indirectly influences membrane potential. This theory is yet to be tested in both larval and imaginal cells.

Shaaya and Sekeris (1970) examined the effects of JH on protocatechuic acid-4-0, β-glucoside synthesis in *Periplaneta americana*. They found that JH acted in the biosynthetic chain leading from tyrosine to protocatechuic acid and that the primary site of biosynthesis of the substance might be the integument. Specific local effects of JH on the integument were described by Levinson *et al.* (1966). Roussel (1970) described the influence of the ca on the heart beat of last instar larvae and adults of *Locusta migratoria*. Allatectomy in the last larval instar slowed down the heart rate, while ca implantation accelerated it. In adults, alltectomy caused a remarkable decrease only if performed after the 8th day. The implantation of a pair of ca did not have any discernible effect on heart-beat. The researcher suggests two possible methods to bring about this effect – the induced release of neurohormones from the cc, or a direct JH action. A third possibility would be that the ca themselves contain an active neurosecretion from the brain. Extirpation of the ca in the last larval instar of the red locust, *Nomadacris septemfasciata*, produces a state comparable to adult diapause, which can be reversed by reimplanting several pairs of ca from normal adults.

Is there more than one corpus allatum hormone ?

The seemingly opposing effects of JH, such as the negative effect on metamorphosis and the positive effect on growth, the activation of larval tissues and, on the other hand, its necessity for the development of such typically imaginal structures as the ovarian follicle cells, led to the hypothesis of two or more ca hormones. A number of indications in favour of this theory have recently been collected. It was debated by Luscher and Springhetti (1960) that the ca in *Kalotermes flavicollis* F. produce two different hormones which have different effects in caste differentiation, a 'juvenile hormone' and 'gonadotropic hormone'. The first of these is supposed to increase the capacity for supplementary reproductive differentiation and the other to initiate pre-soldier differentiation. There is, however, no indication in other insects for the absence of one of the two effects in a gland which both would be present. The practice of this theory, therefore, results in complex contradictions.

Sagesser supported the theory of two ca hormones in his finding that there is no qualitative difference in the oxygen consumption curve between the penultimate and the last larval instar in the cockroach *Leucophaea maderae*. Actually, there is only a quantitative difference in the supposed production of JH, and the qualitative changes in its effects on morphogenesis can be completely explained on this basis. Moreover, there are many other processes acting during each instar, e.g. all those connected with moulting which are practically equivalent in both the penultimate and the last larval instar and which more or less mask the changes because of differences in endocrine balance. All other experimental findings indicate towards the fact that there is only one ca hormone.

According to Schneiderman uses the fact that the extract of the *cecropia* male abdomens does not induce the development of ovaries favours the theory of two ca hormones. This, however, is much more an argument against the identifying the extract with the ca hormone. On the other hand, Wigglesworth (1961) found another morphogenetically active lipoid substance, farnesol, with a positive effect in preventing metamorphosis and in activating ovarian development. Wigglesworth interprets his results as a 'support for the belief that the yolk formation hormone and the juvenile hormone are likely to prove identical'.

There are various arguments in favour of this identity : (1) As confirmed by the histological investigation of the ca during the secretory cycle, there is no evidence for more than one type of secretion, there being only one type of secretory cell in the ca. (2) There is no difference in the endocrine activity of the ca in either direction during the whole life cycle : the ca of larvae from the first to the penultimate instar affect both morphogenesis and ovarian development in the same way as those of the adults. (3) As illustrated elsewhere, there is no major difference in the process of on growth and function of the larval parts of the body and that on the ovarian follicle cells, and both of these effects necessarily result in an increase in oxygen consumption. Even if none of these arguments can be taken as definite evidence of the existence of only one ca hormone, they are, neverthe- less, much more relevant than the opposing arguments. There is therefore no evidence for the hypothesis of two ca hormones.

We get the same result from the fact that all these effects can be reproduced by a single juvenoid (JHa), i.e. by a single chemical substance. Pener (1967) arrived at the same conclusion regarding the identical nature of the ca secretion of both male and female *Schistocerca gregaria* adults. The minute differences which to have been noticed by this author were related to differences between the target organs in the two sexes rather than to differences in actual ca activity.

The control of JH production

Criteria of corpora allata activity

The activity of the ca may be understood either indirectly, by the size and the histological appearance of the gland, or directly, by its effects on the recipient after transplantation. Each of these methods has both advantages as well as disadvantages. There has been a great deal of discussion regarding the reliability of estimating the function of the ca by the increase in the volume of the gland. It must be considered, that there are various instances of increase in volume, both normal and pathological, which are not connected with a corresponding increase in hormone production. For example, the gland increases in size from one instar to the next and this increase is not related with an increase in activity per volume unit. It may be estimated by comparing the size of the gland at two consecutive ecdyses. Similarly, the increase in volume observed following castration in adult females is probably not because of an increase in the secretory activity of the gland but due to restriction in the removal of the secretion by the haemolymph.

The increase in volume owing to the secretory activity of the ca can be calcuted approximately as the difference between the total increase in the volume of the gland and the increase due to growth of the ca up to a given moment.

It may be accepted that under normal conditions in com- parable developmental periods or during short periods, an increase in volume is a definite sign of an increase in the secretory activity of the gland. Histologically, the active stage is characterized by the swollen cytoplasm which stains more deeply with acidophil (nuclear) stains. The amount of cytoplasm increases in proportion to the nucleus even if the nuclei also become swollen. The local or general inhibitory effect on metamorphosis following the implantation of a ca into the last larval instar is usually taken as experimental evidence of JH production. The defect of to this method is that the effect cannot be calculated until after the next ecdysis. During this time the activity of the implanted gland may change drastically.

The probability of a quantitative test for ca activity based on its metabolic effects has been stated by Slama (1963). The implant- ation of an active gland into an allatectomized adult female of *Pyrrhocoris apterus* causes a specific increase in the oxygen consumption of the recipient. This increase appears to be directly proportional to the activity of the gland. Before it can be adopted for general use, however, detailed elaboration of the criteria for the test is required, together with the determination of the time limits of its applicability and the experimental conditions for its full acceptability.

The changes in corpora allata activity during the instar

The confirmation all the experimental (Wigglesworth, 1936, 1940a), histological (Mendes, 1948) and morphological (Novak, 1951b) information agree that the ca are inactive at the beginning of each instar and that the haemolymph does not contain an effective quantity of JH at that time. The amount of hormone gradually increases and the maximum volume of the ca is usually observed in the second half of the intermoult period. The production of the hormone ceases as the moulting process progresses and, as a result, its concentration in the haemolymph decreases. The main reason for this decrease is, undoubtedly, the excretion of JH by the Malpighian tubules as shown by the experiments of Bounhiol (1953). He discovered that ligaturing the Malpighian tubules of the silkworm at the beginning of the last larval instar has an effect similar to ca implantation, i.e. it suppresses metamorphosis to a greater or lesser extent. The operation is effective during a period which corresponds to that for an effective ca implantation. It is longer than the critical periods for AH and MH. The length of this period may, however, vary noticeably in different species. In Lepidoptera, for example, the critical periods for all these hormones seem to be greatly prolonged as a result of the long feeding period and short moulting process. This may explain the discovery by Legay (1948) that the ca of silkworm larvae reach their maximum volume at about the time of each ecdysis. The selective value of the deviation, together with extreme abbreviation of the actual moulting process in many holometabolous insects, evidently depends on the prolongation of the feeding period; this is particularly significantin the last larval instar, where it enables the accumulation of reserve substances for metamorphosis and eventually for ovarian development.

Corpora allata activity during postembryonic development

The production of JH begins in the final stage of embryogenesis after the ca have been fully formed. The secretory activity of the gland continues during larval development with a temporary interruption at the time of each ecdysis, as mentioned above. A particular time is therefore necessary at the beginning of each instar before the active concentration of the hormone is again reached. It is not reached at all in the last larval instar in Hemimetabola, or in the larval and pupal instars in Holometabola. The production of the hormone continues in adult insects. It is clearly cylical in the females of those species which lay eggs in several separate batches, such as cockroaches and many beetles and bugs. No such cycles are noticed in males and in females of those species which lay only one batch of eggs (e.g. *Bombyx mori*) or which lay one egg at a time with wiide gapof time in between the process. (e.g. *Carausius morosus*).

The secretory activity of the corpora allata in the last larval instar

Wigglesworth (1940, 1948) initially believed that JH production ceases completely in the last larval instar or even that the ca actively eliminate JH from the haemolymph during this period. While Piepho (1951), condidered the hypothesis that the presence of some JH in the last larval instar of *Galleria* (Holometabola) was a necessary condition for the development of the pupa : in the complete absence of JH, adult differentiation should occur. This theory was later confiremd by Wigglesworth (1954), Karlson (1956), Schneiderman (1960) and a number of other workers. That the ca continue to produce JH in the last larval instar was shown by Novak and Cervenkova (1959). They implanted ca from last instar nymphs of *Pyrrhocoris apterus* into other last instar nymphs immediately after moulting and at different times during the intermoult period and in most cases they obtained juvenilizing effects of varying extent. This does not prove, however, that the ca were producing a morphogenetically active concentration of JH as assumed by the above hypothesis. There is definite evidence against thisassumption : when the ca are removed from the last larval instar immediately after moulting no morphological change is observed. While, the implantation of another last instar larva ca which itself is inactive in the donor, is sufficient to produce a juvenilizing effect.

Hypertrophy of the corpora allata

Castration of *Calliphora erythrocephala* females immediately after the imaginal moult results in hypertrophy of their ca (Thomsen, E., 1946). The most possible elucidation this phenomenon is that in the absence of the ovaries JH is not removed from the haemolymph. When the concentration of the hormone here, exceeds that inside the gland cells, further diffusion of the hormone into the haemolymph is inhibited. It Therefore, it collect in the gland causing its excess growth and a subsequent decrease in secretory activity. This seems to be a more acceptable explanation than the theory suggested by E. Thomsen (1946, 1947) of a specific hormonal influence of the ovaries which suppresses the activity of the ca. The hypertrophy of the ca observed in functional termite females appears, however, to be of quite a different type, being connected with hyperfunction of the glands.

Effect of an implanted corpus allatum on that of the recipient

The first person to demonstrate at was Pflugfelder (1939a) that the implantation of an active ca causes a more or less complete degene- ration of the recipient's gland. A detailed biometrical analysis of this relationship was carried out by Novak and Rohdendorf (1961) using adult females of *Pyrrhocoris apterus*. They fixed active ca into freshly moulted adults and measured the volume of the recipient's glands on the 3rd, 6th and 9th days after implantation and compared it with that of controls implanted with a piece of muscle of similar size. The ca of the recipients were distinctly smaller than those of the controls. No such effect on the recipient's gland was observed if the implantation was made later in the instar (6th day after moulting), when the recipient's gland had reached its full activity. The explanation of the effect of the implanted gland followes.

The flow of the hormone from the gland into the haemolymph can take occur only by diffusion through the surface of the gland as there is no particular mechanism (e.g. muscular) for this. But diffusion can only take place particular when the concentration of the hormone inside the gland is higher than that in the haemolymph. This condition is not maintained when an active gland is implanted before the recipient;s gland has reached its full activity. Under such circumstances the concentration of the hormone in the haemolymph, produced by the implanted gland, exceeds that inside the recipient's gland and so diffusion and thus also the secretion of the hormone is avoided Of course, if the implantation is effected later in the instar, when the ca of the recipient has reached its full activity, there is already a high concentration of hormone in the haemolymph. This goes beyonds the concentration in the implanted gland so that the position is now reversed and the secretion of the implanted gland is stopped

As mentionted by Engelmann (1965) the ca was first activated by an increase in the blood protein level (to a small degree) and then by maturation. The decrease in ca activity associated with egg maturation can be attributed to a decrease in the haemolymph JH level caused by reduced food intake at this time. In other insects, however, such as the viviparous cockroach *Leucophaea maderae*, it was shown to be due to a hormonal factor produced by the brood sac. JH secretion in Dermaptera was probnable first examined by Ozeki (1965).

External factors controlling the activity of the corpora allata

The direct influence of various factors on JH productio is yet unknown. The only paper analysing the effects of external factors on development is that by Wigglesworth (1952a) on *Rhodnius prolixus*. He discovered that subjecting IVth instar nymphs to a high temperature approaching the living maximum (*c*. 35°C in *Rhodnius*) causes an extension of the intermoult period and the development of somewhat adultoid Vth instars (regressive prothetely). A low temperature approaching the developmental minimum (below 20°C in *Rhodnius)* prolongs the intermoult period more than the high temperature, but the resulting Vth instar nymphs are somewhat juvenile (progressive metathetely). The rsults may be explained as effects on the action of JH in the tissues. A reduced oxygen concentration (below 5°C) produces exactly the same effect even at of high temperature.

A comprehensive morphological investigation of the shape and size of the ca in the bug *Eurygaster integriceps* during postembryonic development and in adults in the different seasons of the year was made by Teplakova (1947). She noticed an increase in the size of the gland in adult females in the spring, associated with the development of the ovaries. The ca of bugs parasitized by larvae of flies of the family Tachinidae were very small. In the summer period of increased activity, the ca of some of the females with functional ovaries were extraordinarily large, whereas in most others they were of medium size. Variations in the size of the gland were also found among females from different heights above sea-level.

Internal factors controlling the activity of the corpora allata

Various explanations have been given for what has been called by Wigglesworth (1948) the 'counting of instars'. Wigglesworth stated on the basis of his experiments with *Rhodnius prolixus* first that the decrease in JH activity in the last larval instar is not connected with the age of the gland. This is obvious from the fact that the ca of the

IVth (penultimate) larval instar when implanted into the Vth (last) instar is able to induce juvenilizing effects, i.e. it retains its activity even when its age is equivalent to that of the last larval instar when it would normally cease to function. From this, Wigglesworth deduced that 'it is not the corpus allatum itself which counts the instars' and he suggested the possibility of a nervous stimulus being the cause of this decline in ca activity in the last larval instar. The good innervation of the gland by the nervus allatus appears to indicate nervous control of the ca.

The theory of nervous control was confirmed by Engelmann and Luscher (1956, 1957) on the basis of their experiments with adult females of *Leucophaea maderae* and further developed by Engelmann (1965). They examined the stimulating effects of an interruption of the nervi corporum cardiacarum on the swelling of the ca previously observed by B. Scharrer (1952). They discoverd that a similar result could be obtained by destroying specific part of the protocerebrum. On the while, no changes were observed after electrocoagulation or surgical removal of the neurosecretory cells of the pars intercerebralis. The treatment induced egg formation in adult females and several extra larval instars in last instar nymphs. The researchers assumed their results as evidence for nervous control of the ca. They believed that the absence of JH both in females carrying oothecae and in last instar larvae is the result of by nervous inhibition. But this is not the only explanation of their results. Deucing from with other known facts, it seems probably that the observed swelling and activation of the ca is because stimulation by the nervous irritation and regeneration process induced by the operation. The fact that no effect was noticed after the removal of the neurosecretory cells may be due to neurosecretory material accumulated in the cc, as ca are incapable of either secretory activity or growth in the absence of AH. A nervous stimulus inhibiting AH production and in this way, undoubtedly, JH production might, however, be the mediator of the inhibitory effect of the oothecae on the activity of the ca, in the same way as the distension of the abdomen has a positive effect on AH production in *Rhodnius* .

A nervous control of the ca was even aceepted by Johansson (1958) in his work with adult *Oncopeltus fasciatus*. He realized that interruption of ca innervation in a starving female can induce egg-laying in the same way as the implantation of an active ca. This treatment, however, was inffective when a brain with unbroken nervi corporum cardiacarum was implanted at the same time as the ca. Further evidence nervous control of ca activity wass provided in the researchpapers by Joly (1945a, b) on *Dytiscus marginalis*, by Detinova (1954) on *Anopheles messae*, by de Wilde (1958) on *Leptinotarsa decemlineata*, etc.

The control of the function of the corpora allata

Over the years there has be lot of research on the question of the innervation of the ca and the form of the supposed nervous and neurosecretory control of their secretory activity. Th relevant papers were reviewed by Highnam (1963). The existence of a fine nerve connecting the ca with the suboesophageal ganglion, demonstrated by Engelmann and Luscher (1956) in *Leucophaea maderae*, was also found in *Periplaneta* in *Schistocerca* and in *Locusta*. Neurosecretory granules were detected in this branch of the nervus allatus. When this nerve is ligatured, neursecretory material accumu- lates on the ca side of the ligature. The diferent forms of control of the activity of the ca have been put formard.

Nervous control, was considered to be of an inhibitory character in some insects, e.g. *Leucophaea* while possibly having a stimulatory effect in others, e.g. in *Oncopeltus* or *Schistocerca* in which removal of the brain neurosecretory cells by cautery or extirpation prevents the ca from attaining their full size. The possible cause of this contradiction was pointed out by Strong (see below). The non-specific effect of regeneration of the operation injury, must also be taken into account. It was discovered, for instance, in *Schistocerca*, that the development of the eggs can be greaty advanced by wounding, as shown by the experiments of Norris (1954) and Highnamn (1961b, 1962a, b).

The function of the nervous system in the activation of the ca was examined by removing parts of the brain and by sectioning the nerves of the glands infemale adults of the Central American locust *(? Schistocerca paranensis* Burm), and by a histological investigation of neurosecretory activity with the following results striking changes in the volume and appearance of the ca were found during maturation, whereas only slight changes occurred in the amounts of neurosecretion at the same time; no material from the median neurosecretory cells was found in the nerves which innervate the ca, but granules from the lateral cells with different histochemical properties from those of the median cells were found in these nerves; extirpation of the cerebral neurosecretory cells resulted in reduced volume and inhibited the activation of the ca. Unlike other experiments, the implantation of ca did not restore oogenesis in allatectomized females. The implanted glands could only become and remain active when they retained intact connections with the central nervous sytem. Never sections and cautery of the neurosecretory centres of the brain showed that the activation of the ca was mediated through the lateral neurosecretory cells in the ipsilateral half of the brain. Transsection of the nervi corporum cardiacarum II resulted in inactivity and reduction of the volume of the ca on the corresponding side. The contradiction between the supposed inhibitory effect of the neurosecretory cells of the brain and their stimulatory effects seems to be due to a disregard of the possible breaking of this connection. It is yet unknown whether the effects of the lateral neurosecretory complex upon the ca are mediated through nervous or neurosecretory stimuli, or, perhaps more probably, through the humoral action of AH.. No function could be attributed to the nervous connections between the ca and the suboesophageal ganglion.

Local effect of neurosecretory granules from the neurosecretory cells of the brain reaching the ca via the nervus allatus. Granules of Gomori-positive materials were found in the ca of various species by a number of authors. B. Scharrer (1958) assumed that 'the restraining influence on the ca takes place by nervous activity, whereas the stimulatory affect is brought about by a substance present in the neurosecretory material'. Khan and Fraser (1962), who found neurosecretory granules in the ca of newly moulted adults of *Periplaneta americana*, believed, contrary of B. Scharrer (1958), that they were responsible for the straining influence there. Mordue (1963) observed neurosecretory material in the ca of female pupae of *Tenebrio moliitor* befor imaginal moulting; after mating, it disappeared from the ca, remaining, however, in the allatic nerves in close proximity to the gland. The ca were found to grow more rapidly in mated than in unmated females. Here again there is no unequivocal evidence of a local physiological function of the neurosecretion in the gland, and on based on the information of neurohormones and their way of reaching the target its explanation is improbable.

Humoral activation by the AH of the brain, corresponding to the activation of the pgl. It seems that activation or inhibition of the production of AH by nervous stimuli from other parts of the body (e.g. the ovaries, ootheca, etc.) or by various external factors (e.g. photoperiodism) might be the only way to control ca activity. Among other findings, this is shown by the fac that the imaginal diapause in *Leptinotarsa*, aphids and *Pyrrhocoris* which is an AH-deficiency syndrome, can be completely abolished by implanting active brains and/or cc, whereas the implantation of active ca is not effective in the absence of AH (in brain and cardiacectomized specimen), as demonstrated by Slama (1964) for *Pyrrhocoris*.

The following result from the available experimental evidence seems to be the most probable. The brain controls the activity of the ca both by neurosecretion (AH) and by nervous activity. The latter may produce either a stimulatory (distension of the abdomen in blood-sucking insects) or an inhibitory effect (carrying of oothecae) depending on to the biology of the species concerned. Neither of these factors, however, seems to be responsible for the decrease in activity of the ca in the last larval instar which is the immediate cause of metamorphosis. The nervous system here is merely a mediator of the external stimuli which on average remain unchanged in each instar.

The dependence of JH production on gland volume

A different view towards the problem of the control of metamorphosis was stated by Kaiser (1949) who stressed on the increasing disproportion between the size of the ca and the pgl and concluded that the amount of JH produced, rapidly decreases compared with that of MH, and the surplus of MH causes metamorphosis. This idea was further developed by L'Helias (1956). He based his views on two suppositions which have been disproved : that of a direct positive effect of MH on metamorphosis; and that of an antagonism between MH and JH. In addition, the actual amount of MH as estimated by the volume of the pgl in relation to the volume of the body, and thus also the concentration of MH in the haemolymph, remains static

Another solutoin which takes into account the surface area and volume of the gland in relation to the volume of the body as responsible for the time of reaching the active concentration of the hormone, was used by the author. The concentration of the hormone in the haemolymph depends on the following factors : (1) the quantity of hormone produced in a unit of time; (2) the volume of the haemolymph and, in direct proportion to this, the volume of the body; (3) the quantity of hormone consumed in the body; (4) the quantity of hormone removed from the haemolymph by the Malpighian tubules (and perhaps other organs) in the unit of time; (5) Another factor controling the rate of release of the hormone is the surface-volume relationship of the gland.

Observation of the changes which these factors undergo during postembryonic development confirms that JH concentration in the haemolymph necessarily decreases with each subsequent instar. Sooner or later, therefore, there comes an instar (which varies with the species) in which an effective minimum JH concentration is not attained before the advancing moulting process inhibits further growth and differentiation. This decides the last larval instar and the beginning of metamorphosis.

The relationship between the surface area and volume of the ca and that of the body, during development

To investigatethe above theory, measurements were made of the surface area and volume of the ca and the volume of the body in freshly hatched larvae (the beginning of the first larval instar) and freshly moulted larvae in the last instar. The results obtained in 18 different species of both Hemimetabola and Holometabola show that there is a marked decrease in the volume of ca relative to that of the body, in addition to a very great decrease in the surface area of the ca compared with the volume of the body. This comparative decrease in the volume of the ca is because of the very slow growth of the gland, which during the whole larval period is slightly greater than that of various parts of the nervous system and much less than the growth of the most other parts of the body, such as the breadth of the head, breadth of the pronotum, length of the tibia, etc. It is confirmed that the relative decrease in the surface area and volume of the ca in all the species studied is too great not to affect the production of JH, and is large enough to explain the lack of an effective concentration of JH in the last larval instar and thus to induce the onset of metamorphosis.

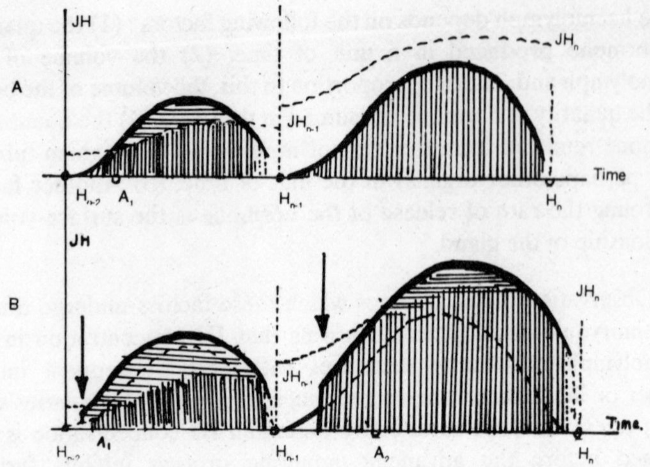

Fig. 7.19. Diagram of the dependence of the type of growth on JH- concentration during the penultimate and last larval instar. A, controls; B, after the implantation of corpora allata (the arrows indicate the moment of implantation). Area with horizontal lines- -proportionate growth, vertical lines – disproportionate (gradient or allometric) growth, Hn—2 – penultimate larval instar, broad line – the actual concentration of JH, ecdysis, Hn—1 – last larval ecdysis; Hn – imaginal ecdysis; dotted line – minimum effective concentration of JH. Abscissa – time, ordinate – JH concentration.

The mode of action of the juvenile hormone

Many theories have been forwarded to explain the mode of action of JH, some obviously contradictory. According to Wiggles- worth's original idea (1936), JH affects morphogenesis on the principle of Goldschmidt's hypothesis of different reaction velocities. Wigglesworth believed that the degree of differentiation attained by a given specimen after ca implantation is the product of competition between two simultaneous processes : the differentiation of the imaginal structures, and moulting. Variation is only possible in the period between the moment of detachment of the old cuticle and that of the deposition of the new. The effect of JH was assumed to depend on accelerating the moulting process so that the permaturely deposited new cuticle would inhibit further differentiation. On this basis, Wigglesworth initially used the term 'inhibitory hormone' for the secretion of the ca. Wigglesworth (1940) himself replaced this theory with the hypothesis of two alternative enzyme systems inside each epidermal cell, larval and imaginal. The larval system was supposed to be dependent upon the presence of JH and the imaginal system in order to operate in the absence of JH. The term 'inhibitory hormone' was therefore changed to 'juvenile hormone' or to 'neotenin'. The hypothesis of polymorphism which has been deseribeed earlier was based on this concept.

While Pflugfelder (1939, 1941) and Weed-Pfeiffer (1945a, b), stressed on the positive effects of JH on total metabolism as shown by their experiments. To avoid the obvious contracdictions, the hypothesis of two different ca hormones was created; one of them being identified with the JH of Wigglesworth, and supposed to affect only larval development. The other, a 'gonadotropic' hormone, was assumed to activate the ovarian follicles and affect metabolism. However no cogent evidence against the first concept Wiggles-worth on a single ca hormone has been produced

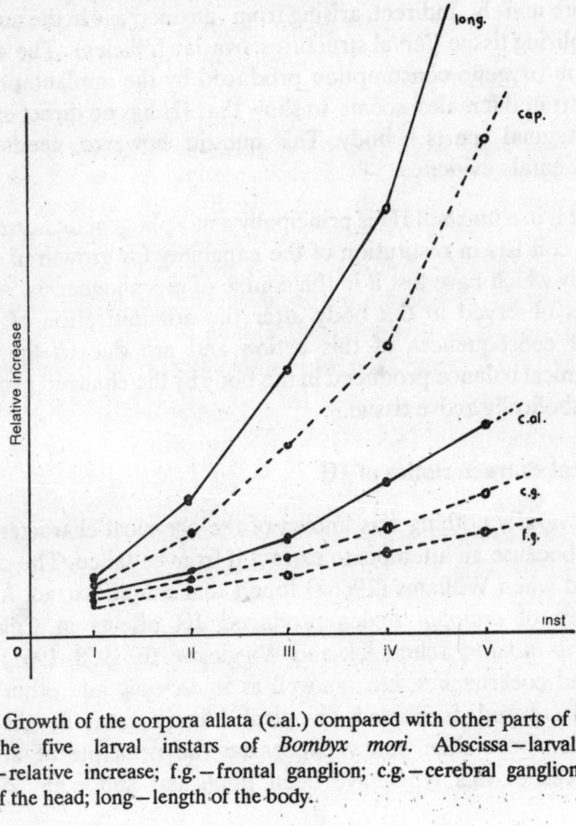

Fig. 7.20. Growth of the corpora allata (c.al.) compared with other parts of the body during the five larval instars of *Bombyx mori*. Abscissa – larval instar, ordinate – relative increase; f.g. – frontal ganglion; c.g. – cerebral ganglion; cap. – breadth of the head; long – length of the body..

On the basis of the gradient-factor as a factor conditioning the JH-independent growth of the imaginal parts of the body, a theory which is discussed later, a hypothesis on the mode of action of JH has been suggested by the author (Novak, 195Ia,b, 1956; cf.p.258). This seems to resolve the above-mentioned contraditions in the experimenatl results and to synthesize earlier vierws. According to this theory, JH produces its effect by taking the place of the GF in those parts of the body which lose it in the course of development. Such parts are the laral parts of the body during larval development, the ovarian follictes in adult femals, and a number of other tissues at particular times during development (cf.p.264. As suggested in the papers by Novalk, Slama, Weing (1961), Novak, Saria (1962) and Slama (1963), the effect of JH seems to depend on the moditioning of protein synthesis as well as other functions in the larval parts of the body. Its effect on oxygen consumption is therefore marely indirect, arising from an increase in the amount of metabolizing tissue (larval structures, ovarian follicles). The negative result on oxygene consumption produced by the implantation of ca into castrated females seems to show that JH has no direct effect on the imagginal parts e body. This questio, however, needs further experimental evidence.

Put it in a nutshell JH is principally a morphogenetic hormone. Its action/ consists in restitution of the capability for growth of parts of the body which have lost it in the course of morphogenesis. All other changes observed in the body after the administration of JH are indirect consequences of this action and are due to the altered biochemical balance produced in the body by the changed proportion of metabolically active tissue.

Chemical characteristics of JH

Previously nothing was known of the chemical character of JH, largely because all attempts to extract it from ca failed. The situation changed when Williams (1966a) found that a lipid extract from the abdomen of *cecropia* males produced JH effects in Coleoptera *(Tenebrio molitor*, Schmialek and Wigglesworth, 1958, 1961) and in bugs and cockroaches, etc., as well as in *cecropia* and other moths. This discovered facilitated chemical identification of the active principle. Since then, many papers on the principle of action of substances of this type have been published and many different chemicals, both natural and synthetic, have been found to produce similar effects. Of the many worke investigated the nature of this '*cecropia* oil', the most successful were Roller and his co-workers who succeeded in isolating the active principle by means of gas chromatography. Later they identified the chemical formula of male *cercopia* extract, which was found to be *methyl-10-epoxy-7-ethyl-3, 11-methyl-236- trideca- dienoate*. A number of related susbtances are also active.

The occurrence of the active principle in the body of the *cecropia* silkworm *(Hyalophora cecropia)* during ontogeny was studied in detail by Gilbert and Schneiderman (1960). As illustreated in the following table, the *cecropia* JH content is fairly high in unfertilized eggs, during the embryonic period and in the freshly hatched larvae. It decreases at the end of larval development and in the pupa, practically disappears when development recommences after the pupal diapause and does not reappear until just before adult emergence, when it increases slowly in the female, but at faster rate in the male.

Roller and Dahm (1970) showed demonatrated that the *cecropia* factor was identical with the actual JH produced by the ca of this species. A number of brain-cc-ca complexes dissected out from pupae two to three days before adult emergence were cultivated in a tissue culture medium for six to seven days. The ability of the complexes to produce JH after this time was illusttrated by reimplanting them into last instar larvae of *Galleria mellonella*. The culture medium was then extracted three times with diethyl ether. With 50 brain-cc-ca complexes, an extract containing 10 000 to 15 000 *Tenebrio* units was obtained. This is evidence that the active principle of *cecropia* oil is actually JH produced by the ca is quite convincing.

TABLE 1

Juvenile hormone content of cecropia during development

Stage of development	*Hormone content per gram fresh weight compared with adult male (%)*
Unfertilized eggs	4.3
7-day-old embryos with yolk	3.7
Ist instar larvae (freshly hatched)	6.4
Vth instar larvae (mixed ages)	0.50
Freshly moulted pupae	0.75
Diapausing pupae (1 month old)	0.55
Chilled pupae (6 months old)	0.0
Pupae 2 days of adult development	0.0
Pupae 11 days of adult development	0.0
Pupae 17 days of adult development	0.0
Pupae 20 days of adult development	0.5
Pupae 22 days of adult development (males)	50.0
Adult males, 2 days	100.0
Adult females, 2 days	3.2